Mometrix
TEST PREPARATION

Common Core High School: Number and Quantity Secrets Study Guide

DEAR FUTURE EXAM SUCCESS STORY

First of all, **THANK YOU** for purchasing Mometrix study materials!

Second, congratulations! You are one of the few determined test-takers who are committed to doing whatever it takes to excel on your exam. **You have come to the right place.** We developed these study materials with one goal in mind: to deliver you the information you need in a format that's concise and easy to use.

In addition to optimizing your guide for the content of the test, we've outlined our recommended steps for breaking down the preparation process into small, attainable goals so you can make sure you stay on track.

We've also analyzed the entire test-taking process, identifying the most common pitfalls and showing how you can overcome them and be ready for any curveball the test throws you.

Standardized testing is one of the biggest obstacles on your road to success, which only increases the importance of doing well in the high-pressure, high-stakes environment of test day. Your results on this test could have a significant impact on your future, and this guide provides the information and practical advice to help you achieve your full potential on test day.

Your success is our success

We would love to hear from you! If you would like to share the story of your exam success or if you have any questions or comments in regard to our products, please contact us at **800-673-8175** or **support@mometrix.com**.

Thanks again for your business and we wish you continued success!

Sincerely,
The Mometrix Test Preparation Team

Need more help? Check out our flashcards at:
http://MometrixFlashcards.com/CCSS

Written and edited by the Mometrix Exam Secrets Test Prep Team
Printed in the United States of America

TABLE OF CONTENTS

Introduction

Thank you for purchasing this resource! You have made the choice to prepare yourself for a test that could have a huge impact on your future, and this guide is designed to help you be fully ready for test day. Obviously, it's important to have a solid understanding of the test material, but you also need to be prepared for the unique environment and stressors of the test, so that you can perform to the best of your abilities.

For this purpose, the first section that appears in this guide is the **Secret Keys**. We've devoted countless hours to meticulously researching what works and what doesn't, and we've boiled down our findings to the five most impactful steps you can take to improve your performance on the test. We start at the beginning with study planning and move through the preparation process, all the way to the testing strategies that will help you get the most out of what you know when you're finally sitting in front of the test.

We recommend that you start preparing for your test as far in advance as possible. However, if you've bought this guide as a last-minute study resource and only have a few days before your test, we recommend that you skip over the first two Secret Keys since they address a long-term study plan.

If you struggle with **test anxiety**, we strongly encourage you to check out our recommendations for how you can overcome it. Test anxiety is a formidable foe, but it can be beaten, and we want to make sure you have the tools you need to defeat it.

Secret Key #1 – Plan Big, Study Small

There's a lot riding on your performance. If you want to ace this test, you're going to need to keep your skills sharp and the material fresh in your mind. You need a plan that lets you review everything you need to know while still fitting in your schedule. We'll break this strategy down into three categories.

Information Organization

Start with the information you already have: the official test outline. From this, you can make a complete list of all the concepts you need to cover before the test. Organize these concepts into groups that can be studied together, and create a list of any related vocabulary you need to learn so you can brush up on any difficult terms. You'll want to keep this vocabulary list handy once you actually start studying since you may need to add to it along the way.

Time Management

Once you have your set of study concepts, decide how to spread them out over the time you have left before the test. Break your study plan into small, clear goals so you have a manageable task for each day and know exactly what you're doing. Then just focus on one small step at a time. When you manage your time this way, you don't need to spend hours at a time studying. Studying a small block of content for a short period each day helps you retain information better and avoid stressing over how much you have left to do. You can relax knowing that you have a plan to cover everything in time. In order for this strategy to be effective though, you have to start studying early and stick to your schedule. Avoid the exhaustion and futility that comes from last-minute cramming!

Study Environment

The environment you study in has a big impact on your learning. Studying in a coffee shop, while probably more enjoyable, is not likely to be as fruitful as studying in a quiet room. It's important to keep distractions to a minimum. You're only planning to study for a short block of time, so make the most of it. Don't pause to check your phone or get up to find a snack. It's also important to **avoid multitasking**. Research has consistently shown that multitasking will make your studying dramatically less effective. Your study area should also be comfortable and well-lit so you don't have the distraction of straining your eyes or sitting on an uncomfortable chair.

The time of day you study is also important. You want to be rested and alert. Don't wait until just before bedtime. Study when you'll be most likely to comprehend and remember. Even better, if you know what time of day your test will be, set that time aside for study. That way your brain will be used to working on that subject at that specific time and you'll have a better chance of recalling information.

Finally, it can be helpful to team up with others who are studying for the same test. Your actual studying should be done in as isolated an environment as possible, but the work of organizing the information and setting up the study plan can be divided up. In between study sessions, you can discuss with your teammates the concepts that you're all studying and quiz each other on the details. Just be sure that your teammates are as serious about the test as you are. If you find that your study time is being replaced with social time, you might need to find a new team.

Secret Key #2 – Make Your Studying Count

You're devoting a lot of time and effort to preparing for this test, so you want to be absolutely certain it will pay off. This means doing more than just reading the content and hoping you can remember it on test day. It's important to make every minute of study count. There are two main areas you can focus on to make your studying count.

Retention

It doesn't matter how much time you study if you can't remember the material. You need to make sure you are retaining the concepts. To check your retention of the information you're learning, try recalling it at later times with minimal prompting. Try carrying around flashcards and glance at one or two from time to time or ask a friend who's also studying for the test to quiz you.

To enhance your retention, look for ways to put the information into practice so that you can apply it rather than simply recalling it. If you're using the information in practical ways, it will be much easier to remember. Similarly, it helps to solidify a concept in your mind if you're not only reading it to yourself but also explaining it to someone else. Ask a friend to let you teach them about a concept you're a little shaky on (or speak aloud to an imaginary audience if necessary). As you try to summarize, define, give examples, and answer your friend's questions, you'll understand the concepts better and they will stay with you longer. Finally, step back for a big picture view and ask yourself how each piece of information fits with the whole subject. When you link the different concepts together and see them working together as a whole, it's easier to remember the individual components.

Finally, practice showing your work on any multi-step problems, even if you're just studying. Writing out each step you take to solve a problem will help solidify the process in your mind, and you'll be more likely to remember it during the test.

Modality

Modality simply refers to the means or method by which you study. Choosing a study modality that fits your own individual learning style is crucial. No two people learn best in exactly the same way, so it's important to know your strengths and use them to your advantage.

For example, if you learn best by visualization, focus on visualizing a concept in your mind and draw an image or a diagram. Try color-coding your notes, illustrating them, or creating symbols that will trigger your mind to recall a learned concept. If you learn best by hearing or discussing information, find a study partner who learns the same way or read aloud to yourself. Think about how to put the information in your own words. Imagine that you are giving a lecture on the topic and record yourself so you can listen to it later.

For any learning style, flashcards can be helpful. Organize the information so you can take advantage of spare moments to review. Underline key words or phrases. Use different colors for different categories. Mnemonic devices (such as creating a short list in which every item starts with the same letter) can also help with retention. Find what works best for you and use it to store the information in your mind most effectively and easily.

Secret Key #3 – Practice the Right Way

Your success on test day depends not only on how many hours you put into preparing, but also on whether you prepared the right way. It's good to check along the way to see if your studying is paying off. One of the most effective ways to do this is by taking practice tests to evaluate your progress. Practice tests are useful because they show exactly where you need to improve. Every time you take a practice test, pay special attention to these three groups of questions:

- The questions you got wrong
- The questions you had to guess on, even if you guessed right
- The questions you found difficult or slow to work through

This will show you exactly what your weak areas are, and where you need to devote more study time. Ask yourself why each of these questions gave you trouble. Was it because you didn't understand the material? Was it because you didn't remember the vocabulary? Do you need more repetitions on this type of question to build speed and confidence? Dig into those questions and figure out how you can strengthen your weak areas as you go back to review the material.

Additionally, many practice tests have a section explaining the answer choices. It can be tempting to read the explanation and think that you now have a good understanding of the concept. However, an explanation likely only covers part of the question's broader context. Even if the explanation makes perfect sense, **go back and investigate** every concept related to the question until you're positive you have a thorough understanding.

As you go along, keep in mind that the practice test is just that: practice. Memorizing these questions and answers will not be very helpful on the actual test because it is unlikely to have any of the same exact questions. If you only know the right answers to the sample questions, you won't be prepared for the real thing. **Study the concepts** until you understand them fully, and then you'll be able to answer any question that shows up on the test.

It's important to wait on the practice tests until you're ready. If you take a test on your first day of study, you may be overwhelmed by the amount of material covered and how much you need to learn. Work up to it gradually.

On test day, you'll need to be prepared for answering questions, managing your time, and using the test-taking strategies you've learned. It's a lot to balance, like a mental marathon that will have a big impact on your future. Like training for a marathon, you'll need to start slowly and work your way up. When test day arrives, you'll be ready.

Start with the strategies you've read in the first two Secret Keys—plan your course and study in the way that works best for you. If you have time, consider using multiple study resources to get different approaches to the same concepts. It can be helpful to see difficult concepts from more than one angle. Then find a good source for practice tests. Many times, the test website will suggest potential study resources or provide sample tests.

Practice Test Strategy

If you're able to find at least three practice tests, we recommend this strategy:

Untimed and Open-Book Practice

Take the first test with no time constraints and with your notes and study guide handy. Take your time and focus on applying the strategies you've learned.

Timed and Open-Book Practice

Take the second practice test open-book as well, but set a timer and practice pacing yourself to finish in time.

Timed and Closed-Book Practice

Take any other practice tests as if it were test day. Set a timer and put away your study materials. Sit at a table or desk in a quiet room, imagine yourself at the testing center, and answer questions as quickly and accurately as possible.

Keep repeating timed and closed-book tests on a regular basis until you run out of practice tests or it's time for the actual test. Your mind will be ready for the schedule and stress of test day, and you'll be able to focus on recalling the material you've learned.

Secret Key #4 – Pace Yourself

Once you're fully prepared for the material on the test, your biggest challenge on test day will be managing your time. Just knowing that the clock is ticking can make you panic even if you have plenty of time left. Work on pacing yourself so you can build confidence against the time constraints of the exam. Pacing is a difficult skill to master, especially in a high-pressure environment, so **practice is vital**.

Set time expectations for your pace based on how much time is available. For example, if a section has 60 questions and the time limit is 30 minutes, you know you have to average 30 seconds or less per question in order to answer them all. Although 30 seconds is the hard limit, set 25 seconds per question as your goal, so you reserve extra time to spend on harder questions. When you budget extra time for the harder questions, you no longer have any reason to stress when those questions take longer to answer.

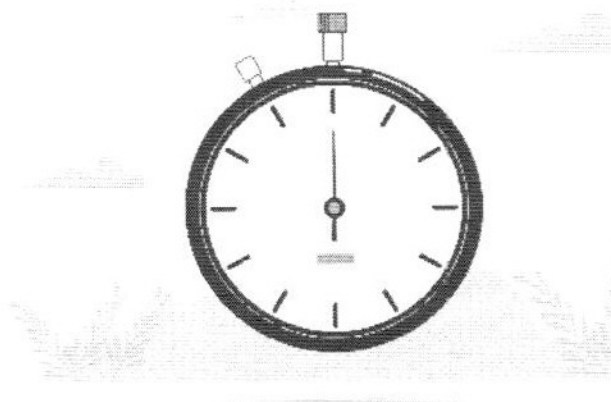

Don't let this time expectation distract you from working through the test at a calm, steady pace, but keep it in mind so you don't spend too much time on any one question. Recognize that taking extra time on one question you don't understand may keep you from answering two that you do understand later in the test. If your time limit for a question is up and you're still not sure of the answer, mark it and move on, and come back to it later if the time and the test format allow. If the testing format doesn't allow you to return to earlier questions, just make an educated guess; then put it out of your mind and move on.

On the easier questions, be careful not to rush. It may seem wise to hurry through them so you have more time for the challenging ones, but it's not worth missing one if you know the concept and just didn't take the time to read the question fully. Work efficiently but make sure you understand the question and have looked at all of the answer choices, since more than one may seem right at first.

Even if you're paying attention to the time, you may find yourself a little behind at some point. You should speed up to get back on track, but do so wisely. Don't panic; just take a few seconds less on each question until you're caught up. Don't guess without thinking, but do look through the answer choices and eliminate any you know are wrong. If you can get down to two choices, it is often worthwhile to guess from those. Once you've chosen an answer, move on and don't dwell on any that you skipped or had to hurry through. If a question was taking too long, chances are it was one of the harder ones, so you weren't as likely to get it right anyway.

On the other hand, if you find yourself getting ahead of schedule, it may be beneficial to slow down a little. The more quickly you work, the more likely you are to make a careless mistake that will affect your score. You've budgeted time for each question, so don't be afraid to spend that time. Practice an efficient but careful pace to get the most out of the time you have.

Secret Key #5 – Have a Plan for Guessing

When you're taking the test, you may find yourself stuck on a question. Some of the answer choices seem better than others, but you don't see the one answer choice that is obviously correct. What do you do?

The scenario described above is very common, yet most test takers have not effectively prepared for it. Developing and practicing a plan for guessing may be one of the single most effective uses of your time as you get ready for the exam.

In developing your plan for guessing, there are three questions to address:

- When should you start the guessing process?
- How should you narrow down the choices?
- Which answer should you choose?

When to Start the Guessing Process

Unless your plan for guessing is to select C every time (which, despite its merits, is not what we recommend), you need to leave yourself enough time to apply your answer elimination strategies. Since you have a limited amount of time for each question, that means that if you're going to give yourself the best shot at guessing correctly, you have to decide quickly whether or not you will guess.

Of course, the best-case scenario is that you don't have to guess at all, so first, see if you can answer the question based on your knowledge of the subject and basic reasoning skills. Focus on the key words in the question and try to jog your memory of related topics. Give yourself a chance to bring the knowledge to mind, but once you realize that you don't have (or you can't access) the knowledge you need to answer the question, it's time to start the guessing process.

It's almost always better to start the guessing process too early than too late. It only takes a few seconds to remember something and answer the question from knowledge. Carefully eliminating wrong answer choices takes longer. Plus, going through the process of eliminating answer choices can actually help jog your memory.

Summary: Start the guessing process as soon as you decide that you can't answer the question based on your knowledge.

How to Narrow Down the Choices

The next chapter in this book (**Test-Taking Strategies**) includes a wide range of strategies for how to approach questions and how to look for answer choices to eliminate. You will definitely want to read those carefully, practice them, and figure out which ones work best for you. Here though, we're going to address a mindset rather than a particular strategy.

Your odds of guessing an answer correctly depend on how many options you are choosing from.

Number of options left	5	4	3	2	1
Odds of guessing correctly	20%	25%	33%	50%	100%

You can see from this chart just how valuable it is to be able to eliminate incorrect answers and make an educated guess, but there are two things that many test takers do that cause them to miss out on the benefits of guessing:

- Accidentally eliminating the correct answer
- Selecting an answer based on an impression

We'll look at the first one here, and the second one in the next section.

To avoid accidentally eliminating the correct answer, we recommend a thought exercise called **the $5 challenge**. In this challenge, you only eliminate an answer choice from contention if you are willing to bet $5 on it being wrong. Why $5? Five dollars is a small but not insignificant amount of money. It's an amount you could afford to lose but wouldn't want to throw away. And while losing $5 once might not hurt too much, doing it twenty times will set you back $100. In the same way, each small decision you make—eliminating a choice here, guessing on a question there—won't by itself impact your score very much, but when you put them all together, they can make a big difference. By holding each answer choice elimination decision to a higher standard, you can reduce the risk of accidentally eliminating the correct answer.

The $5 challenge can also be applied in a positive sense: If you are willing to bet $5 that an answer choice *is* correct, go ahead and mark it as correct.

Summary: Only eliminate an answer choice if you are willing to bet $5 that it is wrong.

Which Answer to Choose

You're taking the test. You've run into a hard question and decided you'll have to guess. You've eliminated all the answer choices you're willing to bet $5 on. Now you have to pick an answer. Why do we even need to talk about this? Why can't you just pick whichever one you feel like when the time comes?

The answer to these questions is that if you don't come into the test with a plan, you'll rely on your impression to select an answer choice, and if you do that, you risk falling into a trap. The test writers know that everyone who takes their test will be guessing on some of the questions, so they intentionally write wrong answer choices to seem plausible. You still have to pick an answer though, and if the wrong answer choices are designed to look right, how can you ever be sure that you're not falling for their trap? The best solution we've found to this dilemma is to take the decision out of your hands entirely. Here is the process we recommend:

Once you've eliminated any choices that you are confident (willing to bet $5) are wrong, select the first remaining choice as your answer.

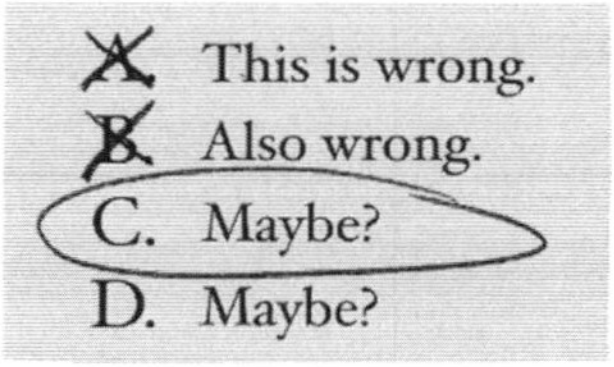

Whether you choose to select the first remaining choice, the second, or the last, the important thing is that you use some preselected standard. Using this approach guarantees that you will not be enticed into selecting an answer choice that looks right, because you are not basing your decision on how the answer choices look.

This is not meant to make you question your knowledge. Instead, it is to help you recognize the difference between your knowledge and your impressions. There's a huge difference between thinking an answer is right because of what you know, and thinking an answer is right because it looks or sounds like it should be right.

Summary: To ensure that your selection is appropriately random, make a predetermined selection from among all answer choices you have not eliminated.

Test-Taking Strategies

This section contains a list of test-taking strategies that you may find helpful as you work through the test. By taking what you know and applying logical thought, you can maximize your chances of answering any question correctly!

It is very important to realize that every question is different and every person is different: no single strategy will work on every question, and no single strategy will work for every person. That's why we've included all of them here, so you can try them out and determine which ones work best for different types of questions and which ones work best for you.

Question Strategies

☑ Read Carefully

Read the question and the answer choices carefully. Don't miss the question because you misread the terms. You have plenty of time to read each question thoroughly and make sure you understand what is being asked. Yet a happy medium must be attained, so don't waste too much time. You must read carefully and efficiently.

☑ Contextual Clues

Look for contextual clues. If the question includes a word you are not familiar with, look at the immediate context for some indication of what the word might mean. Contextual clues can often give you all the information you need to decipher the meaning of an unfamiliar word. Even if you can't determine the meaning, you may be able to narrow down the possibilities enough to make a solid guess at the answer to the question.

☑ Prefixes

If you're having trouble with a word in the question or answer choices, try dissecting it. Take advantage of every clue that the word might include. Prefixes can be a huge help. Usually, they allow you to determine a basic meaning. *Pre-* means before, *post-* means after, *pro-* is positive, *de-* is negative. From prefixes, you can get an idea of the general meaning of the word and try to put it into context.

☑ Hedge Words

Watch out for critical hedge words, such as *likely, may, can, often, almost, mostly, usually, generally, rarely*, and *sometimes*. Question writers insert these hedge phrases to cover every possibility. Often an answer choice will be wrong simply because it leaves no room for exception. Be on guard for answer choices that have definitive words such as *exactly* and *always*.

☑ Switchback Words

Stay alert for *switchbacks*. These are the words and phrases frequently used to alert you to shifts in thought. The most common switchback words are *but, although*, and *however*. Others include *nevertheless, on the other hand, even though, while, in spite of, despite*, and *regardless of*. Switchback words are important to catch because they can change the direction of the question or an answer choice.

☑ FACE VALUE

When in doubt, use common sense. Accept the situation in the problem at face value. Don't read too much into it. These problems will not require you to make wild assumptions. If you have to go beyond creativity and warp time or space in order to have an answer choice fit the question, then you should move on and consider the other answer choices. These are normal problems rooted in reality. The applicable relationship or explanation may not be readily apparent, but it is there for you to figure out. Use your common sense to interpret anything that isn't clear.

Answer Choice Strategies

☑ ANSWER SELECTION

The most thorough way to pick an answer choice is to identify and eliminate wrong answers until only one is left, then confirm it is the correct answer. Sometimes an answer choice may immediately seem right, but be careful. The test writers will usually put more than one reasonable answer choice on each question, so take a second to read all of them and make sure that the other choices are not equally obvious. As long as you have time left, it is better to read every answer choice than to pick the first one that looks right without checking the others.

☑ ANSWER CHOICE FAMILIES

An answer choice family consists of two (in rare cases, three) answer choices that are very similar in construction and cannot all be true at the same time. If you see two answer choices that are direct opposites or parallels, one of them is usually the correct answer. For instance, if one answer choice says that quantity *x* increases and another either says that quantity *x* decreases (opposite) or says that quantity *y* increases (parallel), then those answer choices would fall into the same family. An answer choice that doesn't match the construction of the answer choice family is more likely to be incorrect. Most questions will not have answer choice families, but when they do appear, you should be prepared to recognize them.

☑ ELIMINATE ANSWERS

Eliminate answer choices as soon as you realize they are wrong, but make sure you consider all possibilities. If you are eliminating answer choices and realize that the last one you are left with is also wrong, don't panic. Start over and consider each choice again. There may be something you missed the first time that you will realize on the second pass.

☑ AVOID FACT TRAPS

Don't be distracted by an answer choice that is factually true but doesn't answer the question. You are looking for the choice that answers the question. Stay focused on what the question is asking for so you don't accidentally pick an answer that is true but incorrect. Always go back to the question and make sure the answer choice you've selected actually answers the question and is not merely a true statement.

☑ EXTREME STATEMENTS

In general, you should avoid answers that put forth extreme actions as standard practice or proclaim controversial ideas as established fact. An answer choice that states the "process should be used in certain situations, if…" is much more likely to be correct than one that states the "process should be discontinued completely." The first is a calm rational statement and doesn't even make a definitive, uncompromising stance, using a hedge word *if* to provide wiggle room, whereas the second choice is far more extreme.

⊘ Benchmark

As you read through the answer choices and you come across one that seems to answer the question well, mentally select that answer choice. This is not your final answer, but it's the one that will help you evaluate the other answer choices. The one that you selected is your benchmark or standard for judging each of the other answer choices. Every other answer choice must be compared to your benchmark. That choice is correct until proven otherwise by another answer choice beating it. If you find a better answer, then that one becomes your new benchmark. Once you've decided that no other choice answers the question as well as your benchmark, you have your final answer.

⊘ Predict the Answer

Before you even start looking at the answer choices, it is often best to try to predict the answer. When you come up with the answer on your own, it is easier to avoid distractions and traps because you will know exactly what to look for. The right answer choice is unlikely to be word-for-word what you came up with, but it should be a close match. Even if you are confident that you have the right answer, you should still take the time to read each option before moving on.

General Strategies

⊘ Tough Questions

If you are stumped on a problem or it appears too hard or too difficult, don't waste time. Move on! Remember though, if you can quickly check for obviously incorrect answer choices, your chances of guessing correctly are greatly improved. Before you completely give up, at least try to knock out a couple of possible answers. Eliminate what you can and then guess at the remaining answer choices before moving on.

⊘ Check Your Work

Since you will probably not know every term listed and the answer to every question, it is important that you get credit for the ones that you do know. Don't miss any questions through careless mistakes. If at all possible, try to take a second to look back over your answer selection and make sure you've selected the correct answer choice and haven't made a costly careless mistake (such as marking an answer choice that you didn't mean to mark). This quick double check should more than pay for itself in caught mistakes for the time it costs.

⊘ Pace Yourself

It's easy to be overwhelmed when you're looking at a page full of questions; your mind is confused and full of random thoughts, and the clock is ticking down faster than you would like. Calm down and maintain the pace that you have set for yourself. Especially as you get down to the last few minutes of the test, don't let the small numbers on the clock make you panic. As long as you are on track by monitoring your pace, you are guaranteed to have time for each question.

⊘ Don't Rush

It is very easy to make errors when you are in a hurry. Maintaining a fast pace in answering questions is pointless if it makes you miss questions that you would have gotten right otherwise. Test writers like to include distracting information and wrong answers that seem right. Taking a little extra time to avoid careless mistakes can make all the difference in your test score. Find a pace that allows you to be confident in the answers that you select.

☑ KEEP MOVING

Panicking will not help you pass the test, so do your best to stay calm and keep moving. Taking deep breaths and going through the answer elimination steps you practiced can help to break through a stress barrier and keep your pace.

Final Notes

The combination of a solid foundation of content knowledge and the confidence that comes from practicing your plan for applying that knowledge is the key to maximizing your performance on test day. As your foundation of content knowledge is built up and strengthened, you'll find that the strategies included in this chapter become more and more effective in helping you quickly sift through the distractions and traps of the test to isolate the correct answer.

Now that you're preparing to move forward into the test content chapters of this book, be sure to keep your goal in mind. As you read, think about how you will be able to apply this information on the test. If you've already seen sample questions for the test and you have an idea of the question format and style, try to come up with questions of your own that you can answer based on what you're reading. This will give you valuable practice applying your knowledge in the same ways you can expect to on test day.

Good luck and good studying!

Number and Quantity Overview

REAL, NATURAL, WHOLE, INTEGER, RATIONAL, IRRATIONAL, IMAGINARY, AND COMPLEX NUMBERS

The set of real numbers contains all numbers which have distinct locations on a number line. Natural numbers are real numbers used for counting: 1, 2, 3, 4, ... The set of whole numbers includes the counting numbers along with the number zero: 0, 1, 2, 3, 4, ... The set of integers includes whole numbers and their opposites: ..., -4, -3, -2, -1, 0, 1, 2, 3, 4, ... Rational numbers include any real number which can be expressed as a fraction in which the numerator is an integer and the denominator is a non-zero integer; rational numbers include integers, fractions, terminating, and repeating decimals. Irrational numbers, such as $\sqrt{2}$ and π, are real numbers which are not rational; in decimal form, these numbers are non-repeating and non-terminating, so any decimal (or fractional) representations of irrational numbers are only approximations. The set of imaginary numbers includes all numbers whose squares are negative and therefore excludes any real number; the imaginary number i is defined as the square root of -1. The set of complex numbers encompasses both the real and imaginary; a complex number can be written in the form $a + bi$, where a and b are real numbers, and i is the imaginary number.

RELATIONSHIPS AMONG THESE SETS OF NUMBERS

Below is an example of a diagram which shows that natural numbers are a subset of whole numbers, which are a subset of integers, which are a subset of rational numbers. All of these together with irrational numbers comprise the set of real numbers. Complex numbers include both real and imaginary numbers.

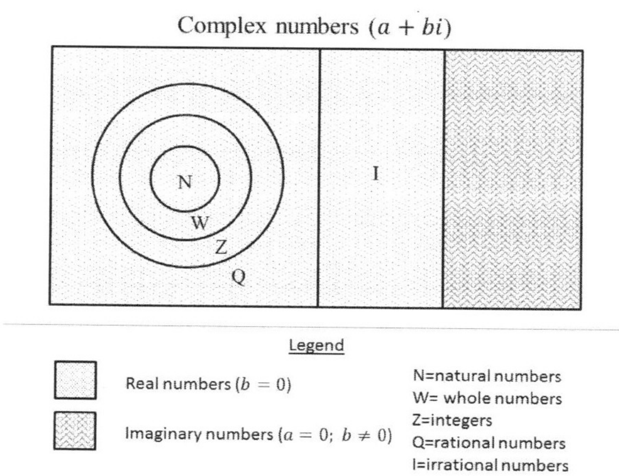

Note that the sizes of the boxes and circles do not reflect the sizes of each set. Each set contains an infinite number of elements.

Problem

Classifying each of the following numbers as a real, natural, whole, integer, rational, irrational, imaginary, and/or complex

1. π

2. -6

3. 22

4. i

5. $\frac{2}{3}$

6. $2 + 3i$

1. π is an irrational number, a real number, and a complex number

2. -6 is an integer, a rational number, a real number, and a complex number

3. 22 is a natural number, a whole number, an integer, a rational number, a real number, and a complex number

4. i is an imaginary number and a complex number

5. $\frac{2}{3}$ is a rational number, a real number, and a complex number

6. $2+3i$ is a complex number

<u>Problem</u>

Determining whether each statement is true

1. All natural numbers are real numbers.

2. Some integers are whole numbers.

3. Imaginary numbers are not real.

4. No integer is irrational.

5. All rational numbers are integers.

1. True; all natural numbers are real numbers.

2. True; some integers are whole numbers. Not all integers, however, are whole numbers, but all whole numbers are integers.

3. True; imaginary numbers are not real.

4. True; no integer is irrational. Integers are part of the set of rational numbers, and irrational numbers are not rational.

5. False; all rational numbers are NOT integers. All integers are rational, but not all rational numbers are integers.

Problem

In the statements below, * and _ represent undefined operations. Name the properties illustrated in each statement and define for which operation(s) $*$ and _ each of these statements is true. Also, identify for which sets of numbers these properties are true

1. $\boldsymbol{a * b = b * a}$

2. $\boldsymbol{a * (b * c) = (a * b) * c}$

3. $\boldsymbol{a * (b\_c) = a * b \_ a * c}$

4. $\boldsymbol{a * b = a}$

1. $\boldsymbol{a * b = b * a}$ **illustrates the commutative property, which is true for any complex numbers** $\boldsymbol{a}$ **and** $\boldsymbol{b}$ **when * represents addition or multiplication.**

2. $\boldsymbol{a * (b * c) = (a * b) * c}$ **illustrates the associative property, which is true for any complex numbers** $\boldsymbol{a}$ **and** $\boldsymbol{b}$ **when * represents addition or multiplication.**

3. $\boldsymbol{a * (b\_c) = a * b \_ a * c}$ **illustrates the distributive property, which is true for any complex numbers a and b when * represents multiplication and _ represents addition or subtraction.**

4. $\boldsymbol{a * b = a}$ **illustrates an identity, where the quantity a is unchanged by the operation * with the quantity b. This property is true for any complex number** $\boldsymbol{a}$ **when * represents addition or subtraction and when** $\boldsymbol{b = 0}$**. This property is true for any complex number** $\boldsymbol{a}$ **when * represents multiplication or division and** $\boldsymbol{b = 1}$**.**

> Note: The operation * can also represent "to the power of" when a is any complex number and $b = 1$ since, by definition, any number raised to the first power is that number.

Zero Exponent and Negative Exponent Properties

The **zero exponent property** states $a^0 = 1$ for all $a \neq 0$. The **negative exponent property** states $a^{-b} = \frac{1}{a^b}$ for all $a \neq 0$.

Consider the series ..., 2, 4, 8, 16, 32, ... Notice that each number is two times the previous number, so the series can be written ... $2, 2 \cdot 2, 2 \cdot 2 \cdot 2, 2 \cdot 2 \cdot 2 \cdot 2, 2 \cdot 2 \cdot 2 \cdot 2 \cdot 2$..., which can be more easily expressed using powers of 2: ..., $2^1, 2^2, 2^3, 2^4, 2^5$, ...

The exponent of each number is one more than the exponent of the previous number. Extend the series a few numbers to the left: ..., $\frac{1}{4}, \frac{1}{2}, 1, 2, 4, 8, 16, 32$, ... Also, apply the pattern of exponents to the series written as powers of two: ..., $2^{-2}, 2^{-1}, 2^0, 2^1, 2^2, 2^3, 2^4, 2^5$, ... Compare the relative positions of newly written terms: $2^0 = 1$; $2^{-1} = \frac{1}{2} = \frac{1}{2^1}$; $2^{-2} = \frac{1}{4} = \frac{1}{2^2}$.

Product of Powers and Quotient of Powers Properties

The **product of powers** property states that when multiplying two monomials with like bases, add together the powers of that base; for example, $x^7 \cdot x^3 = x^{7+3} = x^{10}$ because $(x \cdot x \cdot x \cdot x \cdot x \cdot x \cdot x) \cdot (x \cdot x \cdot x) = x \cdot x \cdot x \cdot x \cdot x \cdot x \cdot x \cdot x \cdot x \cdot x = x^{10}$. The **quotient of powers** property states that when dividing two monomials with like bases, subtract the powers of that base; for example $\frac{x^7}{x^3} = x^{7-3} = x^4$ because $\frac{x \cdot x \cdot x \cdot x \cdot x \cdot x \cdot x}{x \cdot x \cdot x} = \frac{x}{x} \cdot \frac{x}{x} \cdot \frac{x}{x} \cdot x \cdot x \cdot x \cdot x = 1 \cdot 1 \cdot 1 \cdot x \cdot x \cdot x \cdot x = x^4$.

POWER OF A PRODUCT, POWER OF A QUOTIENT, AND POWER OF A POWER

The **power of a product** property states that when multiplying together two monomials with the same power but different bases, multiply the bases and keep the power the same; for example, $x^3 \cdot y^3 = (xy)^3$ because $x^3 \cdot y^3 = x \cdot x \cdot x \cdot y \cdot y \cdot y = x \cdot y \cdot x \cdot y \cdot x \cdot y = (x \cdot y) \cdot (x \cdot y) \cdot (x \cdot y) = (xy)^3$. (Notice the use of the commutative and associate properties of multiplication.) Similarly, the **power of a quotient** property states that when diving two monomials with the same power but different bases, divide the bases and keep the power the same; for example, $\frac{x^3}{y^3} = \left(\frac{x}{y}\right)^3$ because , $\frac{x^3}{y^3} = \frac{x \cdot x \cdot x}{y \cdot y \cdot y} = \frac{x}{y} \cdot \frac{x}{y} \cdot \frac{x}{y} = \left(\frac{x}{y}\right)^3$. Lastly, the **power of a power** property states that when raising a monomial to a power, multiply the power of each term in the monomial by the power to which the monomial is raised; for example, $(x^3y^2)^3 = x^{3\cdot3}y^{2\cdot3} = x^9y^6$ because (x^3 y^2)^3 = $(x \cdot x \cdot x \cdot y \cdot y)\ (x \cdot x \cdot x \cdot y \cdot y)\ (x \cdot x \cdot x \cdot y \cdot y) = x \cdot x \cdot x \cdot x \cdot x \cdot x \cdot x \cdot x \cdot x \cdot y \cdot y \cdot y \cdot y \cdot y \cdot y = x^9y^6$(Again, notice the use of the commutative and associate properties of multiplication.).

Example

Below is an example illustrating $\sqrt[n]{x} = x^{\frac{1}{n}}$.

$\sqrt{3} = 3^{\frac{1}{2}}$	In this example, $x = 3$ and $n = 2$.
$\left(\sqrt{3}\right)^2 = (3^{\frac{1}{2}})^2$	The inverse operation of taking the square root is raising to the second power. Perform the same operation on both sides of the equation.
$3 = 3^{\frac{1}{2}\cdot 2}$	To raise a power to a power, multiply the powers.
$3 = 3^1$	Simplify. $\frac{1}{2} \cdot 2 = 1$.
3=3	Simplify. 3^1 = 3.

Problem

Simplifying the expression $\sqrt[6]{9x^4}$

$(9x^4)^{\frac{1}{6}}$	Rewrite the expression using rational exponents
$(3^2x^4)^{\frac{1}{6}}$	Substitute 3^2 for 9.
$3^{\frac{1}{3}}x^{\frac{2}{3}}$	Use the power of a power property of exponents.
$3^{\frac{1}{3}}x^{\frac{1}{3}}x^{\frac{1}{3}}$	Use the product of a power property to rewrite $x^{\frac{2}{3}}$ as the product of $x^{\frac{1}{3}}$and $x^{\frac{1}{3}}$.
$(3xx)^{\frac{1}{3}}$	Use the product of powers property to rewrite the expression
$\sqrt[3]{3x^2}$	Simplify.

Problem

When two integers are added, subtracted, or multiplied, the resulting sum, difference, or product is also an integer; therefore, the set of integers is closed under these operations. Is the set of rational numbers closed under the same operations?

Given that a, b, c, and d are integers and that b and d do not equal zero,

$$\frac{a}{b}+\frac{c}{d}=\frac{ad+bc}{bd} \text{ and } \frac{a}{b}-\frac{c}{d}=\frac{ad-bc}{bd} \text{ and } \frac{a}{b}\cdot\frac{c}{d}=\frac{ac}{bd}$$

$\frac{a}{b}$ and $\frac{c}{d}$ are rational numbers. The products ad, bc, and bd are integers since the product of two integers is always an integer; likewise, since the sum or difference of two integers is always an integer, $ad+bc$ and $ad-bc$ must also be integers. Therefore, $\frac{ad+bc}{bd}$, $\frac{ad-bc}{bd}$, and $\frac{ac}{bd}$ represent ratios of two integers and are, by definition, rational. So, the set of rational numbers is closed under addition, subtraction, and multiplication.

Problem

Determine if the following statements are true or false. Provide an explanation for the true statement(s) and a counterexample for the false statement(s).

1. The sum of a rational number and an irrational number is always irrational.

2. The sum of two irrational numbers is always irrational

1. It is true that the sum of a rational number and an irrational number is always irrational. Consider for a moment that the statement is false; this would require the existence of some irrational number x that when added to rational number a would produce rational number b: $a+x=b$. Solving the equation for x yields $x=b-a$. The difference of b and a must be rational since the set of rational numbers is closed under subtraction. Since $x=b-a$, x must be a rational number; there is no such irrational number x that when added to a rational number yields a rational number. Therefore, the sum of a rational and an irrational number is always irrational.

2. It is false that the sum of two irrational numbers is always irrational. For example, the sum of irrational numbers $\sqrt{3}$ and $-\sqrt{3}$ is zero, which is rational.

Problem

Determine if the following statements are true or false. Provide an explanation for true statements and a counterexample for false statements

1. The product of a nonzero rational number and an irrational number is always irrational.

2. The product of two irrational numbers is always irrational.

1. It is true that the product of a nonzero rational number and an irrational number is always irrational. Consider for a moment that the statement is false; this would require the existence of some irrational number x that when multiplied by rational number a would produce rational number b: $ax = b$. Solving the equation for x yields $x = \frac{b}{a}; a \neq 0$. If x is equal to the ratio of two rational numbers, it must also, by definition, be rational. (Note that the ratio of two rational numbers can be rewritten as the ratio of two integers.) So, there is no such irrational number x that when multiplied by a rational number yields a rational number. Therefore, the product of a rational and an irrational number is irrational.

2. It is false that the product of two irrational numbers is always irrational. For example, the product of irrational numbers $\sqrt{3}$ and $-\sqrt{3}$ is -3, which is rational.

Problem

Determine if the following statement is always true, sometimes true, or never true: an integer raised to a non-integer power is an integer.

It is **sometimes true** that an integer raised to a non-integer power is an integer. For example, $4^{\frac{1}{2}} = \sqrt{4} = 2$. However, it is not always true. For example, $3^{\frac{1}{2}} = \sqrt{3}$.

Problem

Simplify

1. i^{12}

2. $(2i)^8$

3. $(-i)^{25}$

Notice the pattern that emerges when i is raised to consecutive natural number powers:

$$i^1 = i$$

$$i^2 = -1$$

$$i^3 = i^2 i = (-1)(i) = -i$$

$$i^4 = i^2 i^2 = (-1)(-1) = 1$$

$$i^5 = i^4 i = (1)(i) = i$$

$$i^6 = i^4 i^2 = (1)(-1) = -1$$

$$i^7 = i^4 i^3 = (1)(-i) = -i$$

$$i^8 = i^4 i^4 = (-1)(-1) = 1$$

The imaginary number i raised to any whole power$n$ is always $i, -1, -i$, or 1. To find i^n, either continue writing the pattern $i, -1, -i, 1$ to find the nth number in the pattern, or, more simply, use properties of exponents to rewrite and simplify i^n.

1. $\boldsymbol{i^{12} = (i^4)^3 = (1)^3 = 1}$

2. $\boldsymbol{(2i)^8 = (2^8)(i^4)^2 = (2^4)^2(1)^2 = 16^2 = 256}$

3. $\boldsymbol{(-i)^{25} = (-1)^{25}(i^4)^6(i) = (-1)(1)^6(i) = -i}$

Problem

Determine the values of a and b for which $\sqrt{a} \cdot \sqrt{b} \neq \sqrt{ab}$.

$\sqrt{a} \cdot \sqrt{b} = \sqrt{ab}$ when either a or b is positive or when both a and b are positive. For example,

$\sqrt{25} \cdot \sqrt{4} = 5 \cdot 2 = 10$ and $\sqrt{25 \cdot 4} = \sqrt{100} = 10$, so $\sqrt{25} \cdot \sqrt{4} = \sqrt{25 \cdot 4}$;

$\sqrt{-25} \cdot \sqrt{4} = 5i \cdot 2 = 10i$ and $\sqrt{-25 \cdot 4} = \sqrt{-100} = 10i$, so $\sqrt{-25} \cdot \sqrt{4} = \sqrt{-25 \cdot 4}$.

However, $\sqrt{a} \cdot \sqrt{b} \neq \sqrt{ab}$ when both a and b are both negative. For example,

$\sqrt{-25} \cdot \sqrt{-4} = 5i \cdot 2i = 10i^2 = -10$ but $\sqrt{-25 \cdot -4} = \sqrt{100} = 10$. Since $10 \neq -10$, $\sqrt{-25} \cdot \sqrt{-4} \neq \sqrt{-25 \cdot -4}$.

Problem

Simplify $\sqrt{-200}$.

$\sqrt{-200} = \sqrt{-1 \cdot 2 \cdot 2 \cdot 2 \cdot 5 \cdot 5} = \sqrt{-1} \cdot \sqrt{2^2} \cdot \sqrt{2} \cdot \sqrt{5^2} = i \cdot 2 \cdot \sqrt{2} \cdot 5 = 10i\sqrt{2}$.

Problem

Simplify the following expressions.

1. $\boldsymbol{(2 + 3i) + (6 - 2i)}$

2. $\boldsymbol{(2 + 3i) - (6 - 2i)}$

3. $\boldsymbol{(2 + 3i)(6 - 2i)}$

The commutative, associative, and distributive properties are true for complex numbers. Complex expressions can be simplified just as real variable expressions are simplified. Keep in mind, however, that i is not a variable but is rather the imaginary number, so be sure to simplify i^2 to -1.

1. $\boldsymbol{(2 + 3i) + (6 - 2i) = 8 + i}$

2. $\boldsymbol{(2 + 3i) - (6 - 2i) = -4 + 5i}$

3. $\boldsymbol{(2 + 3i)(6 - 2i) = 12 - 4i + 18i - 6i^2 = 12 + 14i + 6 = 18 + 14i}$

Problem

Simplify $3i(2 + 4i) - (6 + 2i)^2$.

The commutative, associative, and distributive properties are true for complex numbers. Complex expressions can be simplified just as real, variable expressions are simplified. Keep in mind, however, that i is not a variable but is rather the imaginary number, so be sure to simplify i^2 to -1.

$$3i(2 + 4i) - (6 + 2i)^2$$

$$6i + 12i^2 - (36 + 24i + 4i^2)$$

$$6i + 12i^2 - 36 - 24i - 4i^2$$

$$-18i + 8i^2 - 36$$

$$-18i - 8 - 36$$

$$-44 - 18i$$

MODULUS OF A COMPLEX NUMBER

The modulus, or absolute value, of a complex number $z = a + bi$ is the distance from the origin to point z graphed on the complex plane. Since that distance can be represented by the hypotenuse of a right triangle with leg lengths a and b, $|z| = \sqrt{a^2 + b^2}$.

It is **true** that the modulus of a complex number is always a real number and is always positive. Since a and b represent real numbers, their squares are always positive, real numbers. The sum of two positive, real numbers must also be positive and real, so $a^2 + b^2$ is a positive, real quantity. The square root of a positive, real number is also a positive, real number, so $\sqrt{a^2 + b^2}$ must return a number that is both real and positive.

COMPLEX PLANE

The complex plane is created by the intersection of a real, horizontal axis and an imaginary, vertical axis. For a complex number written in the form $a + bi$, a represents the displacement along the real axis and b along the imaginary axis.

Problem

Graph these numbers on the complex plane.

1. -4
2. $2i$
3. $3 - i$

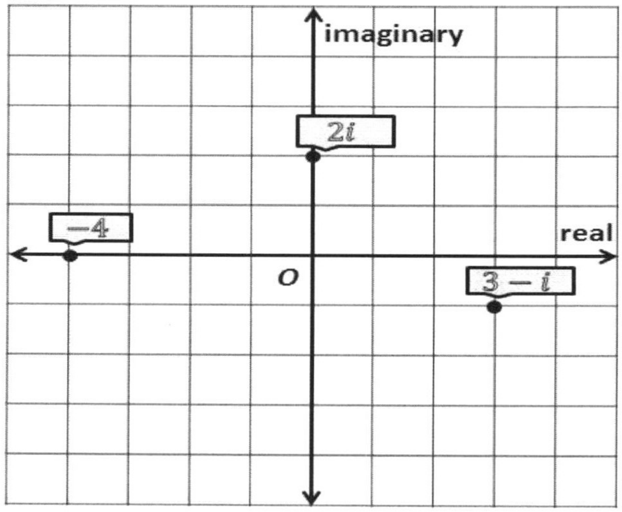

The number -4 is graphed in its appropriate position on the real number line, while $2i$ is graphed on the imaginary axis since its real number component a is 0. The complex number $3 - i$ is represented by a point on the plane which is three units to the right of the origin and one unit down. Another way to represent each of these points graphically on the complex plane is to draw a vector from the origin to the point (a, b). For example, $3 - i$ is graphed below.

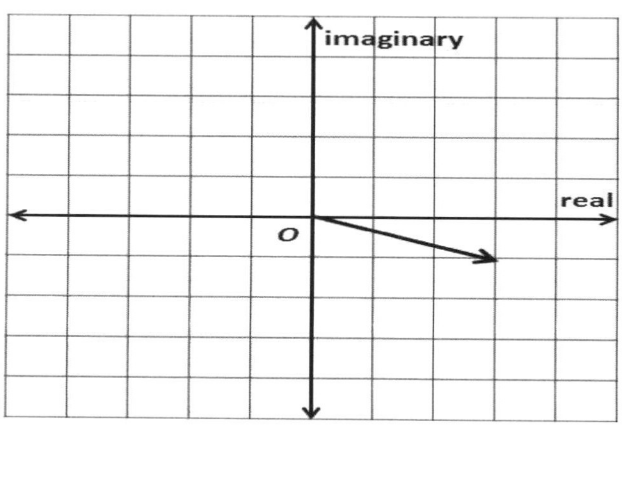

CONJUGATE OF A COMPLEX NUMBER

The conjugate of a complex number $a + bi$ is $a - bi$. It is **true** that the product of a complex number and its conjugate is always real:

$(a + bi)(a - bi) = a^2 - abi + abi - b^2i^2 = a^2 + b^2$. Since a and b are real, and since squares and sums of real numbers are also real, $a^2 + b^2$ is always real.

Problem

Find the modulus and the conjugate of the complex number $4 + 3i$.

The modulus, or absolute value, of a complex number is the number's distance from the origin when graphed on the complex plane. By graphing $4 + 3i$ on the complex plane, it is easy to see that its distance from the origin is the hypotenuse of a right triangle with leg lengths 4 and 3. Recognize 3-4-5 as a common Pythagorean triple, or evaluate and simply the expression $\sqrt{a^2 + b^2}$, where $a = 4$ and $b = 3$: $\sqrt{4^2 + 3^2} = \sqrt{16 + 9} = \sqrt{25} = 5$. So, the modulus of $4 + 3i$ is 5.

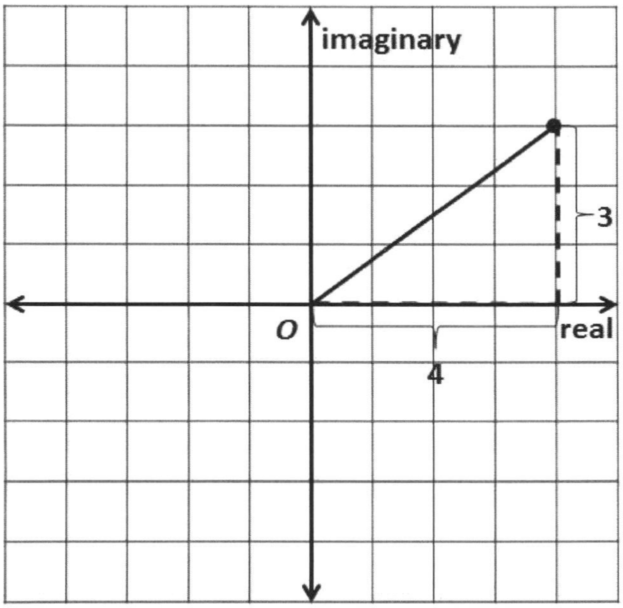

The conjugate of complex number $a + bi$ is defined as $a - bi$, so the conjugate of $4 + 3i$ is $4 - 3i$.

Problem

Write an equation which relates complex number z, its conjugate $\bar{z}$, and its modulus$|z|$.

Let $z = a + bi$. By definition, $\bar{z} = a - bi$, and $|z| = \sqrt{a^2 + b^2}$.

Since $z \cdot \bar{z} = (a + bi)(a - bi) = a^2 - abi + abi - b^2i^2 = a^2 + b^2$, and since

$|z|^2 = \left(\sqrt{a^2 + b^2}\right)^2 = a^2 + b^2$,we have the relationship $\boldsymbol{z \cdot \bar{z} = |z|^2}$.

Use the property $z \cdot \bar{z} = |z|^2$, where z is a complex number and $\bar{z}$ and $|z|$are its conjugate and modulus, respectively, to find the modulus of $3 + 4i$.

When $z = 3 + 4i, \bar{z} = 3 - 4i$.

$$|z|^2 = z \cdot \bar{z} = (3 + 4i)(3 - 4i) = 9 - 16i^2 = 25$$

$$|z| = 5$$

The modulus is represents the distance of a complex number from zero on the complex plane and can therefore not be negative. So, the modulus of $3 + 4i$ is 5.

Problem

Simplify $\frac{2-4i}{1+3i}$.

To simplify the quotient of two complex numbers, multiply the numerator and denominator by the complex conjugate of the denominator.

$$\frac{2-4i}{1+3i} = \frac{2-4i}{1+3i} \cdot \frac{1-3i}{1-3i} = \frac{2-6i-4i+12i^2}{1-9i^2} = \frac{2-10i-12}{1+9} = \frac{-10-10i}{10} = -1 - i$$

Problem

Simplify $\frac{3-i\sqrt{2}}{3+i\sqrt{2}}$.

To simplify the quotient of two complex numbers, multiply the numerator and denominator by the complex conjugate of the denominator.

$$\frac{3-i\sqrt{2}}{3+i\sqrt{2}} = \frac{3-i\sqrt{2}}{3+i\sqrt{2}} \cdot \frac{3-i\sqrt{2}}{3-i\sqrt{2}} = \frac{9-2i\sqrt{2}+2i^2}{9-2i^2} = \frac{9-2i\sqrt{2}-2}{9+2} = \frac{7-2i\sqrt{2}}{11} = \frac{7}{11} - \frac{2\sqrt{2}}{11}i$$

Problem

Use the property $\frac{z_1}{z_2} = \frac{z_1\overline{z_2}}{|z_2|^2}$ to find the quotient of $2 - 4i$ and $1 + 3i$.

$$z_1 = 2 - 4i$$

$$z_2 = 1 + 3i$$

$$\overline{z_2} = 1 - 3i$$

$$|z_2| = \sqrt{1^2 + 3^2} = \sqrt{1+9} = \sqrt{10}$$

So,

$$\frac{2-4i}{1+3i}=\frac{(2-4i)(1-3i)}{\sqrt{10}^2}=\frac{2-6i-4i+12i^2}{10}=\frac{2-10i-12}{10}=\frac{-10-10i}{10}$$
$$=-1-i$$

ARGUMENT OF A COMPLEX NUMBER

A complex number can be represented on the complex plane in both rectangular and polar forms. The polar coordinates (r, θ) can be used to describe the position of a point.

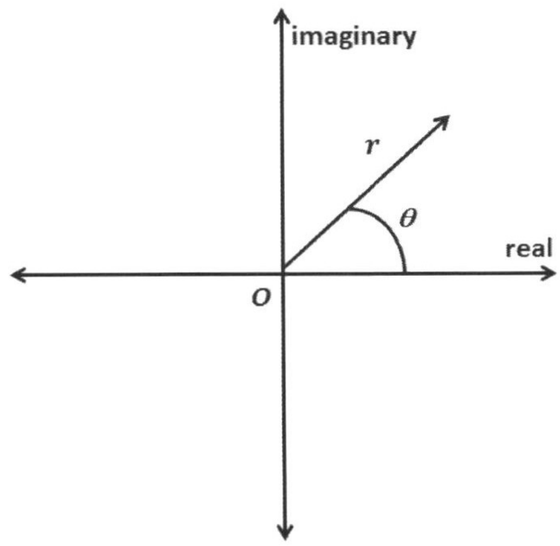

r is the distance from the point to the origin and θ is the angle, in radians, from the positive, real axis to the vector which represents the complex number. This angle θ is called the **argument** of the complex number.

DERIVING THE POLAR FORM OF A COMPLEX NUMBER FROM ITS RECTANGULAR FORM

A complex number can be represented on the complex plane in both rectangular and polar forms. The rectangular coordinates (a, b) or the polar coordinates (r, θ) can be used to describe the position of the same point on the plane.

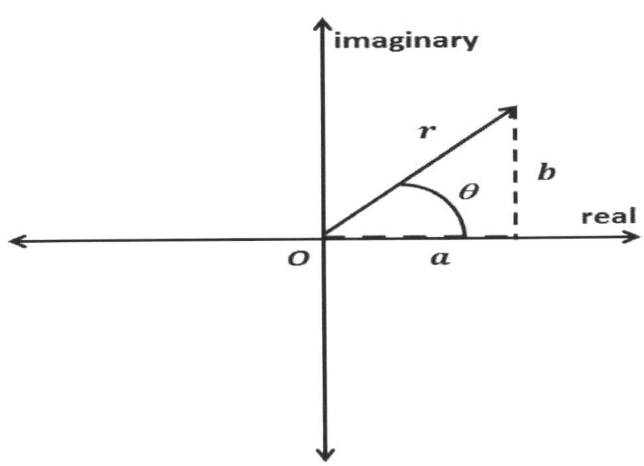

r is the distance from the point to the origin (and for complex z is equivalent to the modulus $|z|$, which equals $\sqrt{a^2+b^2}$), and θ is the angle, in radians, from the positive, real axis to the vector which represents the complex number. To rewrite a complex number $z = a + bi$ in polar form, find

a and b in terms of r and θ. Because $\sin\theta = \frac{b}{r}$ and $\cos\theta = \frac{a}{r}$, we have that $b = r\sin\theta$ and $a = r\cos\theta$. Therefore, $z = r\cos\theta + (r\sin\theta)i$ which is conventionally written $\boldsymbol{z = r(\cos\theta + i\sin\theta)}$.

GRAPHICAL RELATIONSHIP BETWEEN A COMPLEX NUMBER AND ITS CONJUGATE

When graphed on the complex plane, the conjugate of a complex number is the reflection of the complex number across the real axis.

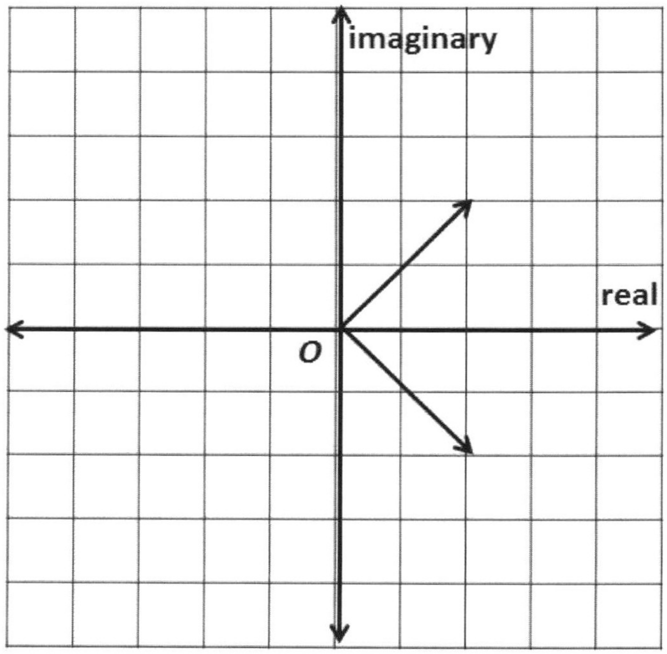

$2 - 2i$ is the reflection across the horizontal axis of $2 + 2i$.

MODULUS OF A COMPLEX NUMBER AND ITS CONJUGATE ARE EQUIVALENT

The complex conjugate of $a + bi$ is $a - bi$. The modulus of a complex number is $\sqrt{a^2 + b^2}$. Substituting –b for b gives $\sqrt{a^2 + (-b)^2}$, which equals $\sqrt{a^2 + b^2}$.

<u>Problem</u>

Write the polar form of these complex number.

$$-3$$

The polar form of a complex number is written in the form $z = r(\cos\theta + i\sin\theta)$, where r is the distance from the point to the origin (and for complex z is equivalent to the modulus $|z|$, which equals $\sqrt{a^2 + b^2}$), and θ is the argument, or the angle, measured in radians, from the positive, real axis to the vector which represents the complex number.

For the real number -3, $r = 3$ and $\theta = \pi$. So, the number can be written in polar form as $3(\cos \pi + i \sin \pi)$

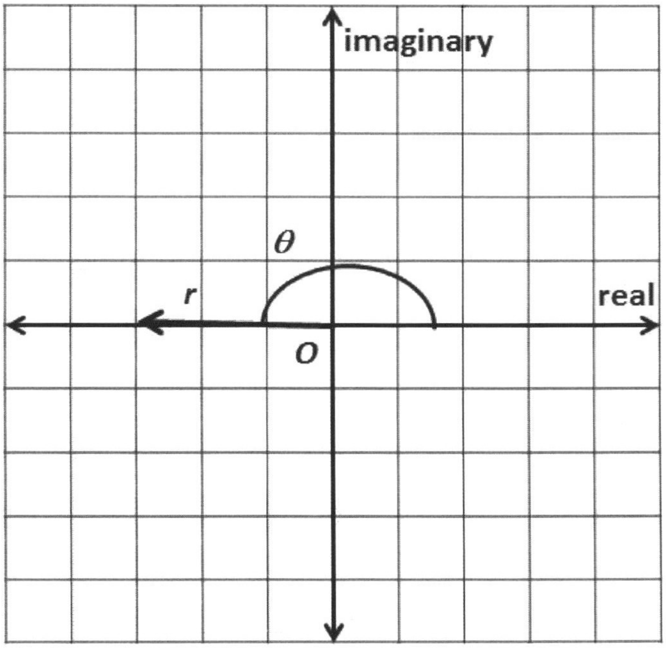

Problem

Write the polar form of these complex number.

$$4i$$

The polar form of a complex number is written in the form $z = r(\cos\theta + i\sin\theta)$, where r is the distance from the point to the origin (and for complex z is equivalent to the modulus $|z|$, which equals $\sqrt{a^2 + b^2}$), and θ is the argument, or the angle, measured in radians, from the positive, real axis to the vector which represents the complex number.

For the imaginary number $4i$, $r = 4$ and $\theta = \frac{\pi}{2}$. So, the number can be written in polar form as $4(\cos\frac{\pi}{2} + i\sin\frac{\pi}{2})$.

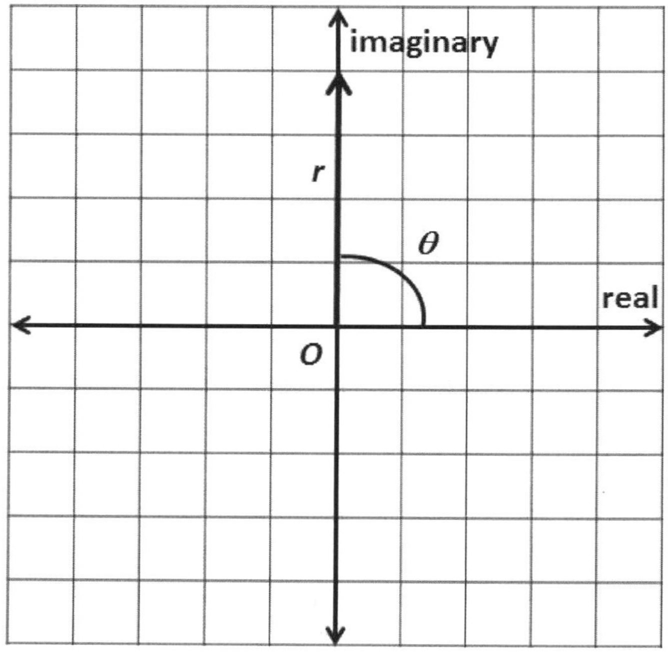

Problem

Write $2 + 2i$ in polar form and graph it on the complex plane.

The polar form of a complex number is written in the form $z = r(\cos\theta + i\sin\theta)$, where r is the distance from the point to the origin (and for complex z is equivalent to the modulus $|z|$, which equals $\sqrt{a^2 + b^2}$), and θ is the argument, or the angle, measured in radians, from the positive, real axis to the vector which represents the complex number.

Calculate r and θ for complex number $2 + 2i$.

$$r = \sqrt{a^2 + b^2} = \sqrt{2^2 + 2^2} = \sqrt{8} = 2\sqrt{2}.$$

$$\theta = \tan\left(\frac{2}{2}\right)$$

$$\theta = \frac{\pi}{4}$$

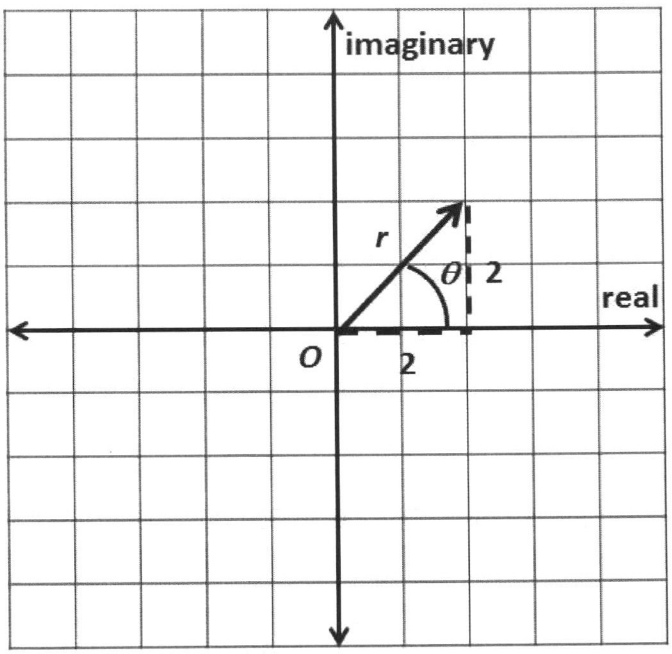

So, the number can be written in polar form as $2\sqrt{2}(\cos\frac{\pi}{4} + i\sin\frac{\pi}{4})$.

MULTIPLYING TWO COMPLEX NUMBERS WRITTEN IN POLAR FORM

A complex number z can be written in polar form as $z = r(\cos\theta + i\sin\theta)$, where $r = |z| = \sqrt{a^2 + b^2}$ and θ is the argument of z. Consider the two complex numbers $z_1 = |z_1|(\cos\theta_1 + i\sin\theta_1)$ and $z_2 = |z_2|(\cos\theta_2 + i\sin\theta_2)$.

$$z_1 z_2 = |z_1|(\cos\theta_1 + i\sin\theta_1)\,|z_2|(\cos\theta_2 + i\sin\theta_2)$$
$$= |z_1||z_2|(\cos\theta_1 + i\sin\theta_1)(\cos\theta_2 + i\sin\theta_2)$$
$$= |z_1||z_2|(\cos\theta_1\cos\theta_2 + i\cos\theta_1\sin\theta_2 + i\sin\theta_1\cos\theta_2 + i^2\sin\theta_2\sin\theta_2)$$
$$= |z_1||z_2|(\cos\theta_1\cos\theta_2 + i(\cos\theta_1\sin\theta_2 + \sin\theta_1\cos\theta_2) - \sin\theta_2\sin\theta_2)$$
$$= |z_1||z_2|\big((\cos\theta_1\cos\theta_2 - \sin\theta_2\sin\theta_2) + i(\cos\theta_1\sin\theta_2 + \sin\theta_1\cos\theta_2)\big)$$

Use the trigonometric identities $\cos(A + B) = (\cos A\cos B - \sin A\sin B)$ and $\sin(A + B) = (\cos A\sin B + \sin A\cos B)$ to simplify the expression in parentheses, which gives $\boldsymbol{z_1 z_2 = |z_1||z_2|\big((\cos(\theta_1 + \theta_2) + i\sin(\theta_1 + \theta_2)\big)}$.

Notice that the complex number which results from multiplying z_1 and z_2 has a modulus of $\boldsymbol{|z_1||z_2|}$and an argument of $\theta_1 + \theta_2$. So, to multiply two complex numbers, multiply their moduli and add their arguments.

<u>Problem</u>

> Rewrite $\sqrt{3} - i$ in polar form. Then, using the properties of multiplication for complex numbers in polar form, calculate $\left(\sqrt{3} - i\right)^3$.

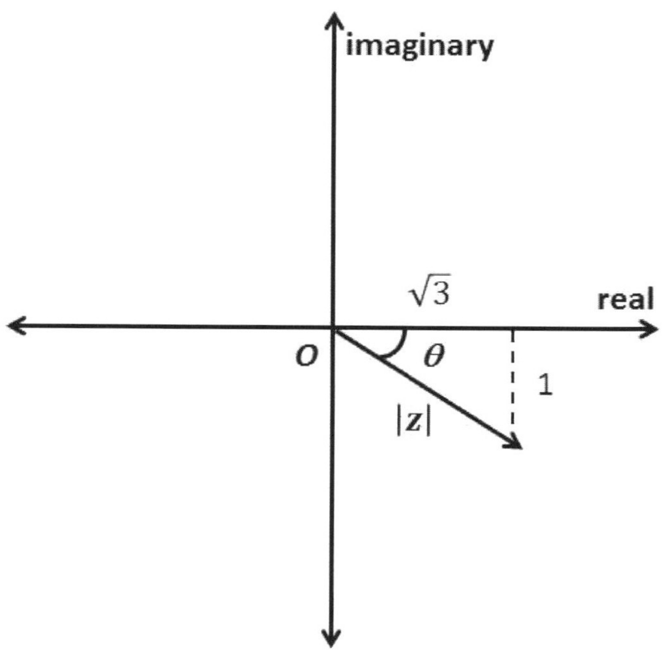

First find the modulus and argument of the complex number $\sqrt{3} - i$. Sketch the graph of $\sqrt{3} - i$.

$|z| = \sqrt{a^2 + b^2} = \sqrt{3 + 1} = 2.$

$\sin\theta = -\frac{1}{2}$ and $\cos\theta = \frac{\sqrt{3}}{2}$, so $\theta = -\frac{\pi}{6}$.

We can rewrite $-\frac{\pi}{6}$ as positive angle $\frac{11\pi}{6}$.

The complex number $\sqrt{3} - i$ can be written in polar form as $z = 2(\cos\frac{11\pi}{6} + i\sin\frac{11\pi}{6})$.

Using the property of multiplication for complex numbers, z^3 has a modulus of $2 \cdot 2 \cdot 2 = 8$ and an argument of $(\frac{11\pi}{6}) + (\frac{11\pi}{6}) + (\frac{11\pi}{6}) = \frac{33\pi}{6}$, which is coterminal with $\frac{3\pi}{2}$ because $\frac{33\pi}{6} - 2\pi - 2\pi = \frac{3\pi}{2}$.

So, $z^3 = 8\left(\cos\left(\frac{3\pi}{2}\right) + i\sin\left(\frac{3\pi}{2}\right)\right) = 8(0 + i(-1)) = 8(-i) = -8i.$

Problem

Write the complex number $4\sqrt{2}(\cos\frac{5\pi}{4} + i\sin\frac{5\pi}{4})$ in rectangular form.

First, sketch the graph of $4\sqrt{2}(\cos\frac{5\pi}{4} + i\sin\frac{5\pi}{4})$.

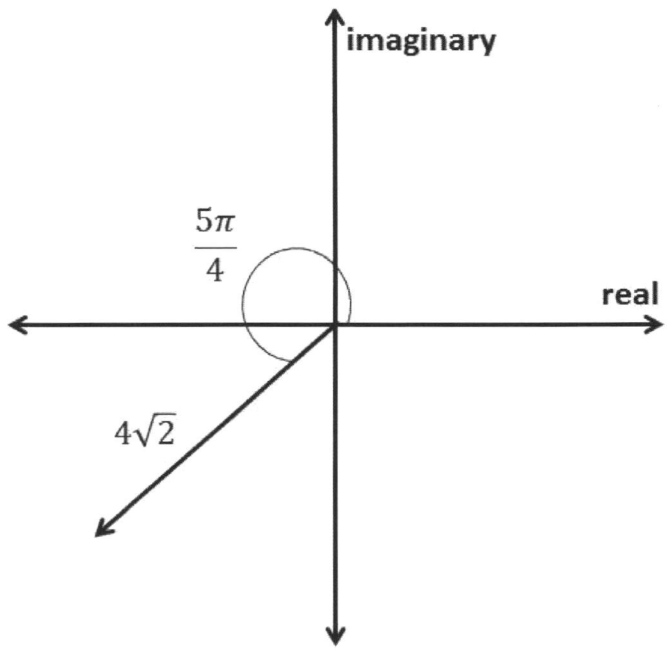

$\frac{b}{4\sqrt{2}} = \sin\frac{5\pi}{4} = \frac{\sqrt{2}}{2}$ simplifies to $b = -4$

$\frac{a}{4\sqrt{2}} = \cos\frac{5\pi}{4} = -\frac{\sqrt{2}}{2}$ simplifies to $a = -4$.

Since the complex number is in the third quadrant, both its real and imaginary components are negative, so its rectangular form is $-4 - 4i$.

Finding the Roots of a Quadratic Polynomial

One way to solve a quadratic equation in the form $ax^2 + bx + c = 0$, where a≠0, is to find the zeros, or x-intercepts, of the related quadratic function's **graph**. Because the roots may be irrational or complex, this method is not the best one to choose when an exact solution is needed.

A second way to solve a quadratic equation in the form $ax^2 + bx + c = 0$, where a≠0 is to **factor**; according to the Zero Product Property, a solution to the equation exists when either factor is equal to zero. This method is especially useful when $c = 0$ or when the factors of the trinomial are easy to determine.

A third way to solve a quadratic equation is to **use the Square Root Property**, completing the square if necessary. (To use this method, the quadratic expression must be a perfect square but need not be set equal to zero.) This method can be used to find both real and complex roots.

A fourth way to solve a quadratic equation in the form $ax^2 + bx + c = 0$ is to use the Quadratic Formula, $x = \frac{-b\pm\sqrt{b^2-4ac}}{2a}$. This method can also be used to find real and complex solutions.

Using the Discriminant of a Quadratic Equation to Determine the Number and Types of Solutions to the Equation

A quadratic equation in the form $ax^2 + bx + c = 0$, where a≠0, has a discriminant of $b^2 - 4ac$. If the discriminant is greater than zero, the equation has two real solutions; more specifically, if the discriminant is a perfect square, those solutions are rational, but if the discriminant is not a perfect square, the solutions are irrational. If the discriminant is equal to zero, the equation has one real, rational root (and is called a double root.). If the discriminant is less than zero, the equation has two complex solutions, which are a conjugate pair.

Problem

Factor $100x^2 + 9$ using complex factors.

Rewrite $100x^2 + 9$ as $100x^2 - 9i^2$. This is a difference of squares, which can be factored as $(10x - 3i)(10x + 3i)$.

Problem

Factor $2x^2 + 8i - 8$ using complex factors.

First, factor out the common factor, which is 2. This gives $2x^2 + 8i - 8 = 2(x^2 + 4i - 4)$. Then, rewrite $x^2 + 4i - 4$ as $x^2 + 4i + 4i^2$, which is a perfect square and can therefore be factored as $(x + 2i)^2$. So, $2x^2 + 8i - 8 = 2(x + 2i)^2$.

Fundamental Theorem of Algebra

The Fundamental Theorem of Algebra states that a single variable polynomial of nth degree with $n > 0$ has *at least* one root. Furthermore, the polynomial has *exactly n* roots if multiplicities, or roots which appear more than once, are counted. In other words, a polynomial of nth degree can always be written as the product of a non-zero constant and *n* linear factors.

For example, any quadratic polynomial can be written as $k(x - r_1)(x - r_2)$, where r_1r and r_2 are complex roots and k is a non-zero constant. Keep in mind that the set of complex numbers includes real numbers, so the roots may not always be real numbers.

Example

Solve $x^2 + 10x + 25 = 16$ by

1. Graphing

2. Factoring

To solve a quadratic equation by graphing or factoring, rewrite it in the form $ax^2 + bx + c = 0$.

$$x^2 + 10x + 25 = 16$$

$$x^2 + 10x + 25 - \mathbf{16} = 16 - \mathbf{16}$$

$$x^2 + 10x + 9 = 0$$

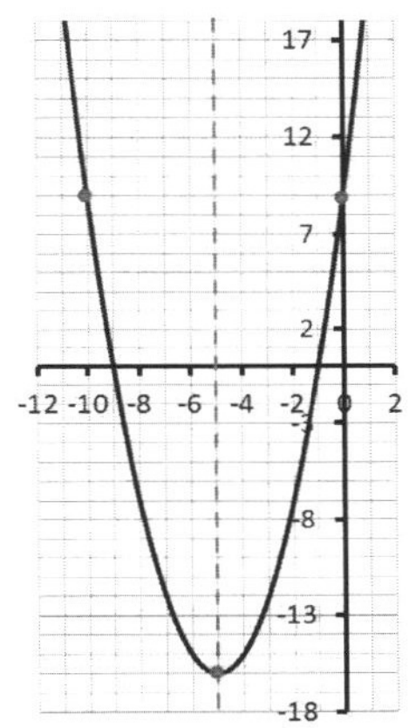

1. Graph the equation $y = x^2 + 10x + 9$. The graph of a quadratic equation is a parabola. To find the parabola's line of symmetry, use the equation $x = -\frac{b}{2a}$.

$$x = -\frac{10}{2(1)} \rightarrow x = -5$$

The vertex of the parabola lies on the line of symmetry. Find y when $x = -5$.

$$y = (-5)^2 + 10(-5) + 9 = 25 - 50 + 9 = -16$$

The vertex is $(-5, -16)$.

Graph the vertex and line of symmetry. Find and graph another point on the parabola and reflect it across the line of symmetry.

$$y = (0)^2 + 10(0) + 9 = 9$$

Another point on the curve is $(0,9)$, which reflects to $(-10,9)$.

The roots of the equation are the x –intercepts -1 and -9.

2. Factor and solve using the Zero Product Property

$$(x+9)(x+1)=0$$

$$(x+9)=0 \qquad (x+1)=0$$

$$x=-9 \qquad x=-1$$

Example

Solve $x^2+10x+25=16$ by using

1. The Square Root Property 2. The quadratic formula

1. The trinomial on the left side of the equation is a perfect square. Solve using the Square Root Property.

$$(x+5)^2=16$$

$$\sqrt{(x+5)^2}=\pm\sqrt{16}$$

$$x+5=\pm 4$$

$$x=-9 \qquad x=-1$$

2. Rewrite the equation in the form $ax^2+bx+c=0$ to use the quadratic formula,
$x=\frac{-b\pm\sqrt{b^2-4ac}}{2a}$.

$$x^2+10x+25=16$$

$$x^2+10x+25-\mathbf{16}=16-\mathbf{16}$$

$$x^2+10x+9=0$$

$$x=\frac{-10\pm\sqrt{10^2-4(1)(9)}}{2(1)}=\frac{-10\pm\sqrt{100-36}}{2}=\frac{-10\pm\sqrt{64}}{2}=\frac{-10\pm 8}{2}$$

$$x=-\frac{2}{2}=-1 \quad x=-\frac{18}{2}=-9$$

Example

Solve $5x^2+120=0$

One way to solve this equation is to use the Square Root Property.

$$5x^2+120=0$$

$$5x^2=-120$$

$$\frac{5x^2}{5}=\frac{-120}{5}$$

$$x^2=-24$$

$$\sqrt{x^2} = \pm\sqrt{-24}$$

$$x = \pm 2i\sqrt{6}$$

Example

Solve $3x^2 - 6x = -4$.

Rewrite the equation in the form $ax^2 + bx + c = 0$ and use the quadratic formula, $x = \frac{-b \pm \sqrt{b^2 - 4ac}}{2a}$, to find the roots.

$$3x^2 - 6x + 4 = 0$$

$$x = \frac{-(-6) \pm \sqrt{(-6)^2 - 4(3)(4)}}{2(3)}$$

$$x = \frac{6 \pm \sqrt{36 - 48}}{6}$$

$$x = \frac{6 \pm \sqrt{-12}}{6}$$

$$x = \frac{6 \pm 2i\sqrt{3}}{6}$$

$$x = 1 \pm \frac{i}{3}\sqrt{3}$$

Notice that the quadratic polynomial has two complex roots, which are conjugate pairs.

Example

Sketch the graph of a quadratic function with

1. No real roots
2. One real root
3. Two real roots

The graph of a quadratic function with no real roots does not intersect the x-axis. An example of such a graph is

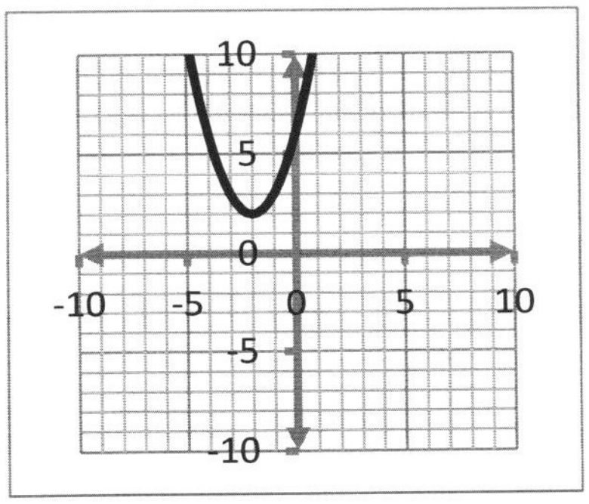

The graph of a quadratic function with one real root intersects the x-axis at one point; in other words, the vertex of the parabolic graph lies on the x-axis. An example of such a graph is

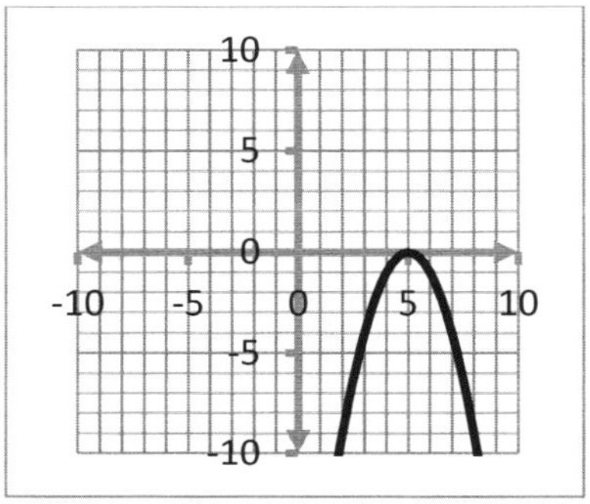

The graph of a quadratic function with two real roots intersects the x-axis at two points. An example of such a graph is

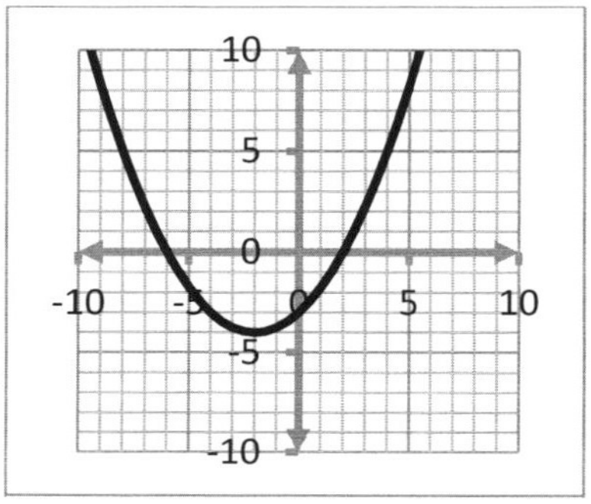

Problem

Determine whether the following statement is always true, sometimes true, or never true: the quadratic equation $ax^2 + bx + c = 0$ has two complex roots when *a, b,* and *c* are real numbers.

According to the Fundamental Theorem of Algebra, $ax^2 + bx + c = 0$ has two roots. These roots are sometimes real, and sometimes they are not. Because the set of real numbers is a subset of the set of complex numbers, the quadratic equation $ax^2 + bx + c = 0$, ALWAYS has two complex roots, regardless of the values of a, b and c.

Scalar Quantity vs. Vector Quantity

A scalar quantity can be defined by magnitude alone. Examples of scalar quantities are many and include mass and temperature because a magnitude, or single number, is sufficient to describe each of these quantities. For example, the mass of a paperclip is about 1g, and the boiling point of water is 100°C under standard temperature and pressure. It would make no sense to describe the mass of a paperclip as 1g upward or the boiling point of water as 100°C to the southwest.

Vector quantities, on the other hand, can only be fully described using both magnitude and direction. An example of a vector quantity is force; for example, gravity exerts a downward (towards the center of the earth) pull, and its magnitude depends on the mass of the object.

DISTANCE AND DISPLACEMENT VS. SPEED AND VELOCITY

1. The distance an object travels is a scalar quantity, which means it is reported with a single unit of magnitude, whereas displacement is a vector quantity, which is reported with both magnitude and direction.

2. Similarly, speed is a scalar quantity, while velocity is a vector quantity. The speed an object travels is the travel distance divided by the travel time, whereas the object's velocity is the object's displacement divided by time. Thus, velocity has both magnitude and direction

Example

A girl walks partly around a city block to get to the library. After traveling 50 m east, then 200 m north, then 50 m west, she reaches her destination in five minutes. Find the distance traveled, displacement, average walking speed, and velocity.

The distance the girl travels is $50 \text{ m} + 200 \text{ m} + 50 \text{ m} = 300 \text{ m}$. However, her final position with respect to her original position, or her displacement, is 200 m north. Her average speed is $\frac{300 \text{ m}}{5 \text{ min}} = \frac{60 \text{ m}}{\text{min}}$, and her velocity is $\frac{200 \text{ m}}{5 \text{ min}} = \frac{40 \text{ m}}{\text{min}}$ northward.

Problem

Define these symbols in terms of vectors and scalars:

1,v

2. $\|v\|$

3.c

1. A vector can be named as a letter written in boldface type, such as v.

2. $\|v\|$ indicates the magnitude of vector v.

3. A non-boldface letter indicates a scalar, which has magnitude by no direction.

Problem

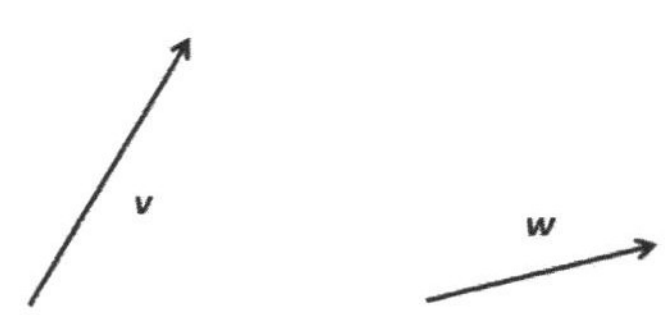

Using a diagram, show how to add vectors v and w by

1. Adding end-to-end;

2. Using the parallelogram rule.

1. Place the tail of vector $\boldsymbol{w}$ at the tip of vector $\boldsymbol{v}$. The vector which connects the ta

il of vector $\boldsymbol{v}$ to the tip of vector $\boldsymbol{w}$ represents the sum of vectors $\boldsymbol{v}$ and $\boldsymbol{w}$. Notice that vector addition is commutative: $\boldsymbol{v} + \boldsymbol{w} = \boldsymbol{w} + \boldsymbol{v}$.

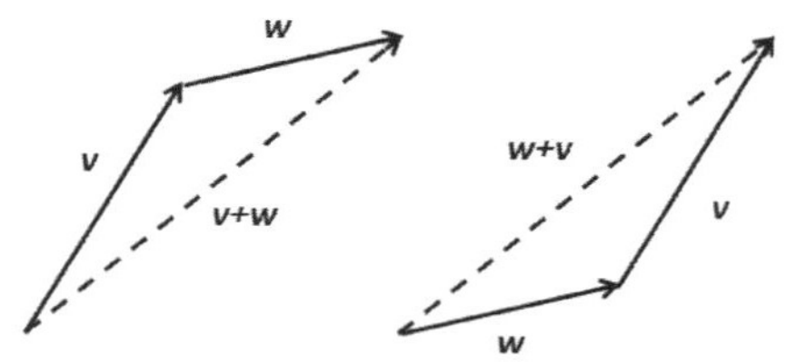

2. Place the tails of vectors ***v*** and ***w*** together. Construct a parallelogram. The diagonal vector which has its tail at the tails of ***v*** and ***w*** represents the sum of

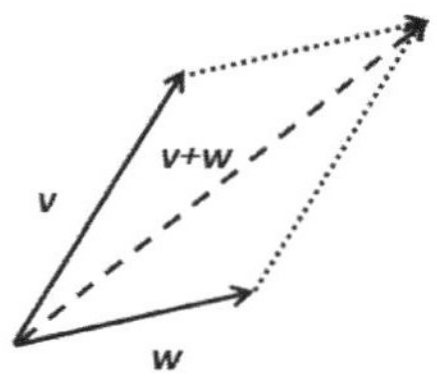

vectors ***v*** and ***w***.

Problem

Use the parallelogram rule for vector addition to graphically represent the sum of the complex numbers $z = 2 + 3i$ and $w = -4 + 2i$ on the complex plane.

Represent each complex number $a + bi$ as a vector on the complex plane by drawing a ray from the origin to the point located a units along the horizontal axis and b units along the vertical axis. Construct a parallelogram. The diagonal ray originating from the origin represents the sum of the two complex numbers.

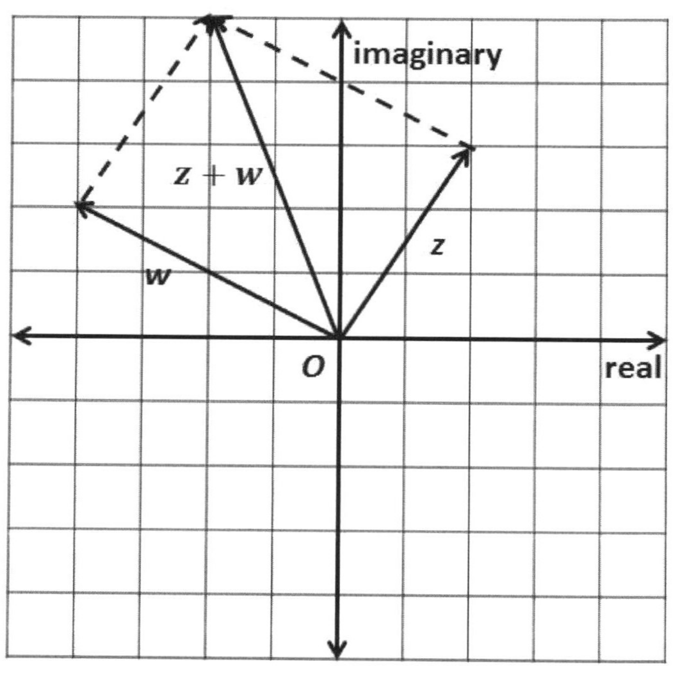

$$z + w = -2 + 5i$$

Problem

Use a vector diagram to illustrate each scenario: Find the resultant velocity of the plane with respect to the ground.

- A plane travels at an air speed of 150 mph amidst a tailwind of 30 mph.
- A plane travels at an air speed of 150 mph amidst a headwind of 30 mph.

A plane travels at an air speed of 150 mph amidst a tail wind of 30 mph. The magnitude of plane's velocity vector is five times the magnitude of the wind's velocity vector, and the direction of both vectors is the same. Using head to tail addition, the resultant velocity is 180 mph.

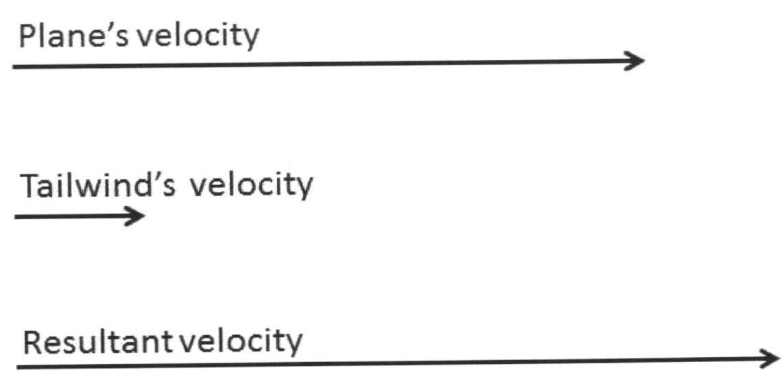

A plane travels at an air speed of 150 mph amidst a head wind of 30 mph. The magnitude of plane's velocity vector is five times the magnitude of the wind's velocity vector, and the vectors point in opposite directions. Using head to tail addition, the resultant velocity is 120 mph.

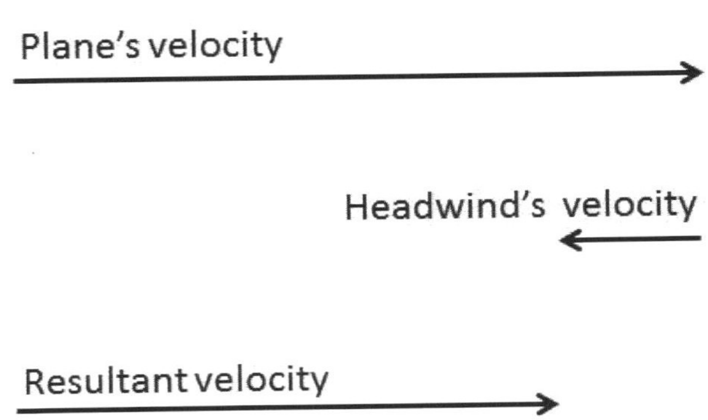

Example

Given vectors v and w, draw the resultant vector $v - 2w$.

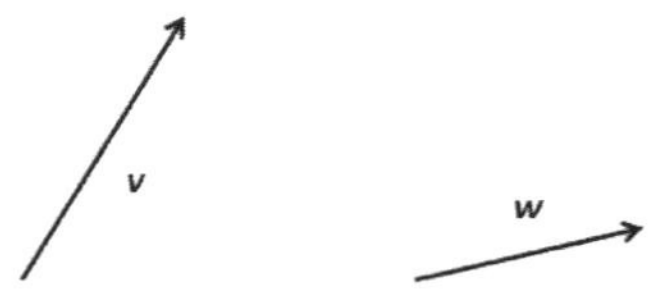

First, draw the vector represented by $2\boldsymbol{w}$. Multiplying vector $\boldsymbol{w}$ by 2 is the same as adding vector $\boldsymbol{w}$ to itself; the result is a vector which has the same direction as vector $\boldsymbol{w}$ but twice its magnitude. Subtracting vector $2\boldsymbol{w}$ from vector $\boldsymbol{v}$ is the same as adding vector -2**w** to vector $\boldsymbol{v}$, that is, $\boldsymbol{v} - 2\boldsymbol{w} = \boldsymbol{v} + (-2\boldsymbol{w})$. Vector $-2\boldsymbol{w}$ has the same magnitude but opposite direction as vector $2\boldsymbol{w}$.

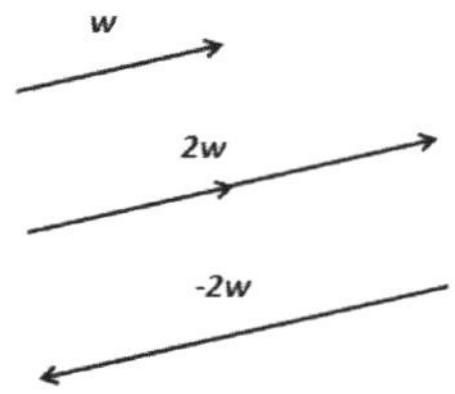

Next, place the tail of vector $-2\boldsymbol{w}$ at the tip of vector $\boldsymbol{v}$. The vector which connects the tail of vector v and the head of vector $-2\boldsymbol{w}$ represents $\boldsymbol{v} - 2\boldsymbol{w}$.

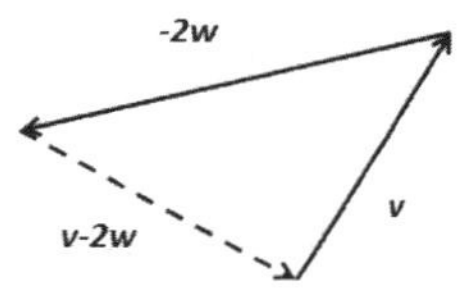

<u>Example</u>

Graphically represent the difference of the complex numbers $z = 2 + 3i$ and $w = -4 + 2i$ on the complex plane.

Represent $z = 2 + 3i$ as a vector on the complex plane by drawing a ray from the origin to the point located 2 units along the horizontal axis and 3 units along the vertical axis. Then, represent $-w = 4 - 2i$ on the complex plane by drawing a ray from the origin to the point located 4 units along the horizontal axis and -2 units along the vertical axis. Construct a parallelogram. The diagonal ray originating from the origin represents the sum of the z and $-w$. Since $z + (-w) = z - w$, this vector represents the difference of z and w. See that $z - w = 6 + i$.

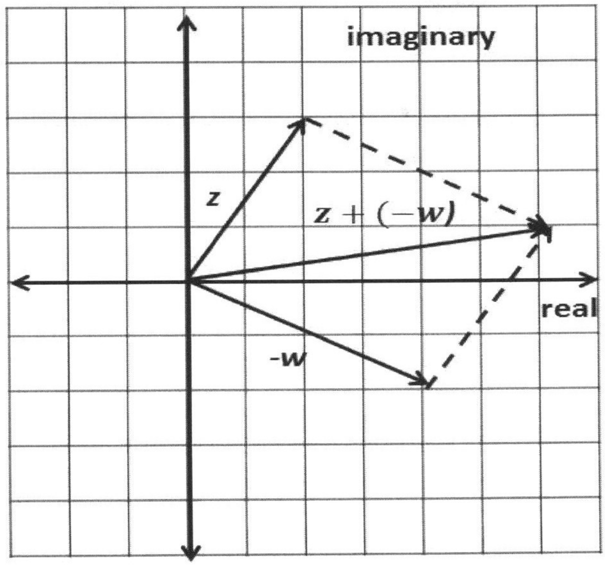

Example

Show that the distance between two numbers on the complex plane is equal to the modulus of the difference of those two numbers.

On the left, two complex numbers are represented on the complex plane as vectors ***v*** and ***w***, and the distance between the two is shown with a dotted line. The graph on the right depicts the difference between vectors ***v*** and ***w***: the difference of vector **v** and **w** is equal to the sum of vectors ***v*** and **-w**, which has the same magnitude but opposite direction of **w**. The sum of two vectors can be found using the parallelogram rule as shown. Notice that the length of the line which represents the distance between ***w*** and ***v*** is the same as the magnitude of the vector which represents the difference of ***v*** and ***w***. The modulus of a complex number is the same as the magnitude of a vector which represents the complex number on the complex plane.

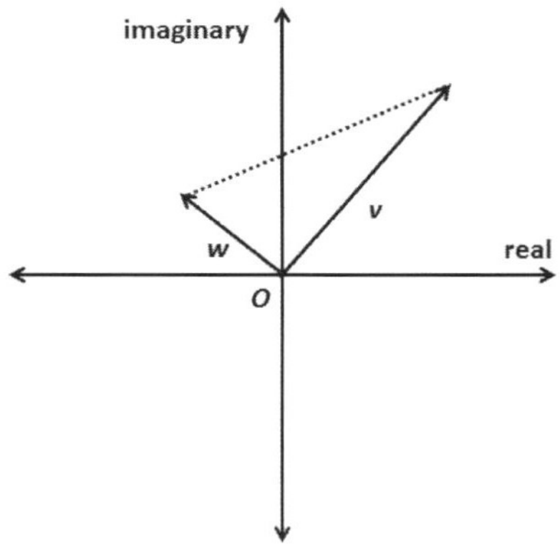

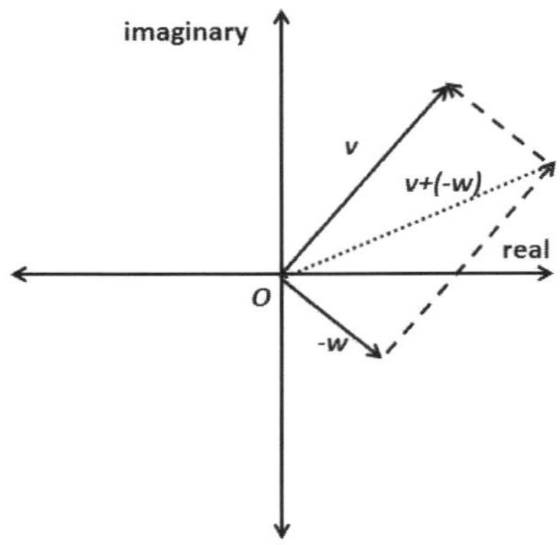

<u>Problem</u>

Find the distance between complex numbers $z = 2 + 3i$ and $w = -4 + 2i$ on the complex plane.

The distance between two numbers on the complex plane is equal to the modulus of the difference of those two numbers. So, find the difference of z and w:

$$z - w = 2 + 3i - (-4 + 2i) = 6 + i.$$

Then, find the modulus of the complex number $6 + i$.

$$|6 + i| = \sqrt{a^2 + b^2} = \sqrt{6^2 + 1^2} = \sqrt{37}.$$

Problem

Find the midpoint of a line segment graphed on the complex plane

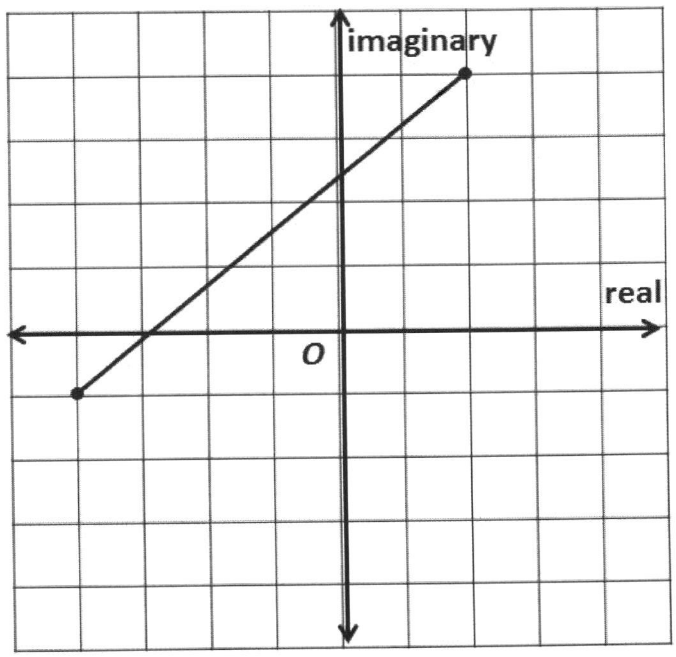

The two endpoints of the line segment are $-4 - i$ and $2 + 4i$. The midpoint can be found by averaging the two endpoints:

$$\frac{-4 - i + (2 + 4i)}{2} = \frac{-2 + 3i}{2} = -1 + \frac{3}{2}i$$

VECTOR COMPONENTS AND VECTOR RESOLUTION

Vectors which are themselves neither vertical nor horizontal have vertical and horizontal components. For example, consider vector ***v***, showing a northeasterly displacement. The dashed lines show the northward and eastward components, the combined influence of which is equivalent to the single northeastward vector ***v***. A vector ***v*** which originates at the origin has components(v_1, v_2), where v_1 and v_2 are the x- and y-coordinates at the tip, or terminal point, of the vector.

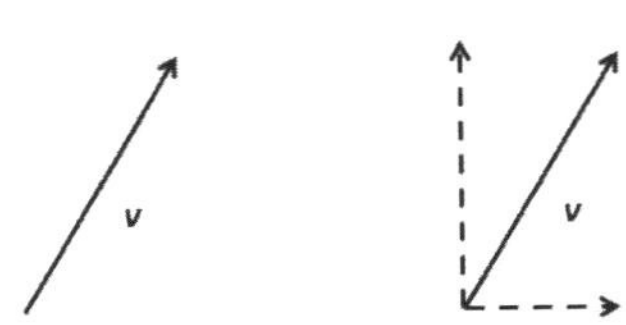

Vector resolution is the process of determining the magnitudes of the component vectors. One method of vector resolution uses the trigonometry. Sketch a right triangle with two-dimensional vector $\boldsymbol{v}$ as the hypotenuse and its components as the legs.

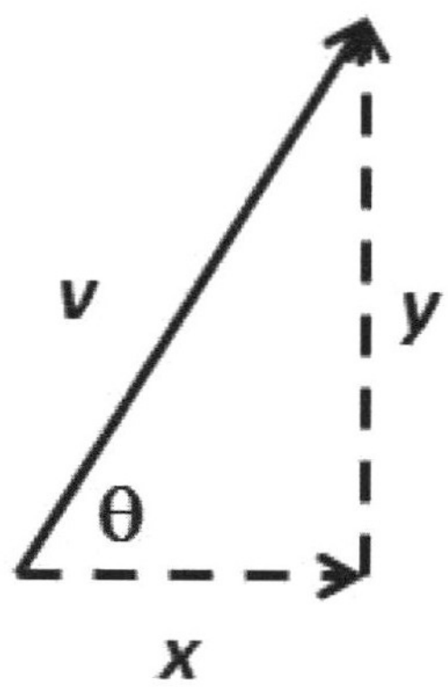

Since $\sin\theta = \frac{y}{v}$, the magnitude of component vector $y = v\, sin\, \theta$; since $\cos\theta = \frac{x}{v}$, the magnitude of component vector $x = v\cos\theta$.

<u>Example</u>

> Find the components of a vector with an initial point of $(2, 3)$ and a terminal point of $(4, 7)$.
>
> The components of a vector can be found by subtracting its initial point from its terminal point: $(4 - 2, 7 - 3) = (2, 4)$.

Example

Find the components of vector *v*.

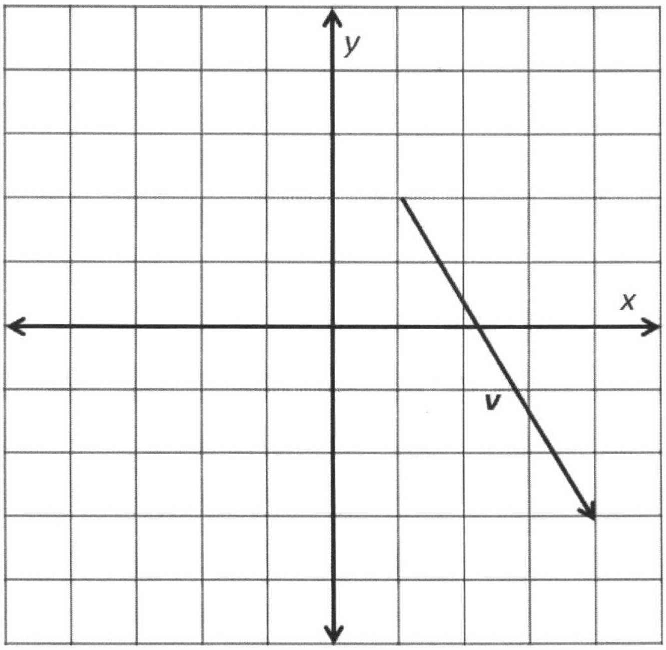

One way to find the components of a vector is to subtract its initial point from its terminal point. The initial point of vector $\boldsymbol{v}$ is $(1, 2)$, and its terminal point is (4,-3). So we have $(4 - 1, -3 - 2) = (3, -5)$.

Problem

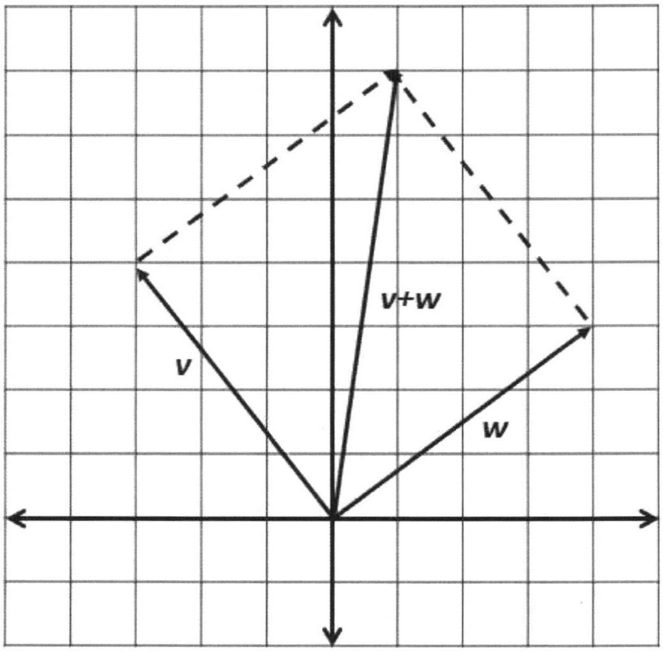

Determine whether or not the following statement is true: the magnitude of the sum of two vectors is equal to the sum of the two vectors' magnitudes. If true, provide an explanation, and if false, provide a counterexample.

False; the magnitude of the sum of two vectors is NOT always equal to the sum of the two vectors' magnitudes.

In this counterexample, both vectors $\boldsymbol{v}$ and $\boldsymbol{w}$ have a magnitude of 5, so the sum of their magnitudes is 10. The magnitude of $\boldsymbol{v} + \boldsymbol{w}$, however, is $\sqrt{1^2 + 7^2} = \sqrt{50} = 5\sqrt{2}$.

Example

For vector $v = (4, 3)$ and vector $w = (-3, 4)$,

1. Use component-wise addition to find $v + w$;

2. Use component-wise scalar multiplication to find $2w$.

1. The sum of two vectors is equal to the sum of their components. Using component-wise addition, $v + w = (4 + (-3), 3 + 4) = (1, 7)$.

2. To multiply a vector by a scalar, multiply each component by that scalar. Using component-wise scalar multiplication, $2w = 2(-3, 4) = (2 \cdot -3, 2 \cdot 4) = (-6, 8)$.

Example

For vector v, which represents a downward force of 10 N, find the direction and magnitude of vector cv when

1. $c = 3$

2. $c = -2$

$||cv||=|c|v$

In other words, the magnitude of vector is equal to the product of the absolute value of scalar c and the magnitude of vector $\boldsymbol{v}$.

When $c > 0$, the direction of vector cv is the same as the direction of vector $\boldsymbol{v}$; when $c < 0$, the direction of vector cv is the opposite as the direction of vector $\boldsymbol{v}$.

1. $||cv|| = |c|||v|| = |3|10 = 30$.

Since c is positive, cv has a direction along $\boldsymbol{v}$.
So, cv represents a downward force of 30 N when $c = 3$.

2. $||cv||=|c|||v|| = |-2|10 = 20$

Since c is negative, cv has a direction against $\boldsymbol{v}$.

So, cv represents a upward force of 20 N when $c = -2$.

Problem

A girl's rowing speed in still water is 4 mph. To cross a river, she rows east, perpendicular to the current, which is 3 mph to the north. Draw a vector diagram to illustrate this scenario, and find magnitude and direction of the resultant vector.

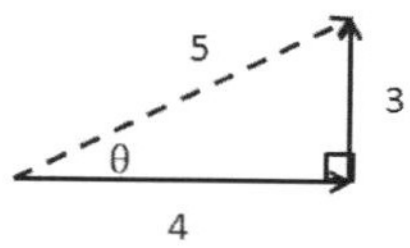

The vectors form a 3-4-5 right triangle, so the magnitude of the resultant vector is 5 mph.

$$\tan\theta = \frac{3}{4} \qquad \theta = 36.9°$$

The direction of the resultant vector is 36.9° north of east.

Problem

A plane traveling at an airspeed of 300 mph due south encounters a 50mph wind from the southeast. Determine the velocity of the plane relative to the ground.

A wind from the southeast blows to the northwest. Place together the tails of vector representing wind velocity and the vector representing the tail velocity and use the parallelogram rule to show the resultant velocity. To calculate the magnitude and direction of the resultant velocity, determine the northerly and westerly components of the wind.

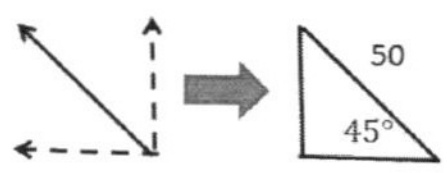

$$\sin 45 = \frac{N}{50} \qquad N = 35.4$$

$$\cos 45 = \frac{W}{50} \qquad W = 35.4$$

Since the wind has a northerly component of 35.4 mph, the southerly component of the resultant velocity is the plane's airspeed, 300 mph, minus 35.4 mph, which is 264.6mph. Use this value and the wind's easterly component to find the magnitude and direction of the plane's velocity with respect to the ground.

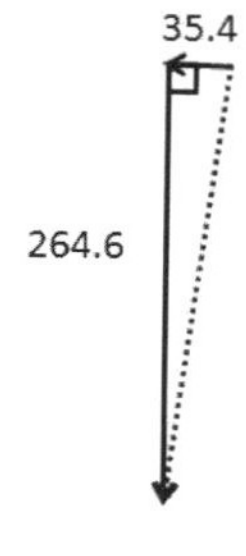

The magnitude of the vector is $\sqrt{35.4^2 + 264.6^2} = 267$

The angle south of west is

$\tan\theta = \frac{264.6}{35.4}$ $\quad\theta = 82.4°$.

So, the plane's velocity relative to the ground is 267 mph, 82.4° south of west.

Matrix

A matrix is a rectangular array of elements arranged in rows and columns, which are usually enclosed by brackets. A matrix can be described by its dimensions, its shape, or its elements:

- A matrix with m rows and n columns is an $m \times n$ matrix
- A matrix with the same number of rows and columns is a square matrix; a matrix consisting of only one row is a row matrix; and a matrix consisting of only one column is a column matrix.
- A matrix in which every element is zero is a zero matrix, which can have any dimension.

Example

Solve $\begin{bmatrix} 2x+y \\ x-y \end{bmatrix} = \begin{bmatrix} 7 \\ -1 \end{bmatrix}$ for x and y.

Two matrices are equal when their corresponding elements are equal. Use this property of matrices to write a system of equations:

$$2x + y = 7$$

$$x - y = -1$$

Solve the system using graphing, substitution, or elimination. Eliminating y gives:

$$2x + y = 7$$

$$\underline{x - y = -1}$$

$$3x = 6$$

$$\frac{3x}{3} = \frac{6}{3}$$

$$x = 2$$

Substitute 2 for x in either equation to find y.

$$3 - y = -1$$

$$3 - y - \mathbf{2} = -1 - \mathbf{2}$$

$$-y = -3$$

$$y = 3$$

Multiply a Matrix by a Scalar

To multiply a matrix by a scalar, or constant, multiply each element in the matrix by that constant.

$$k\begin{bmatrix} a & b \\ c & d \\ e & f \end{bmatrix} = \begin{bmatrix} ka & kb \\ kc & kd \\ ke & kf \end{bmatrix}$$

Example

Organize the vertices $A = (-1, 4)$, $B = (-3, -1)$, and $C = (0, 5)$of ΔABC into a matrix. Write a second matrix which represents the vertices ofΔXYZ if ΔXYZ is a dilation of $\Delta ABC$with three times its perimeter.

One way to organize the vertices is into a 2×3 matrix in which the top row represents the x-coordinates and the bottom row represents the y-coordinates of vertices A, B, and C, respectively.

$$\begin{bmatrix} -1 & -3 & 0 \\ 4 & -1 & 5 \end{bmatrix}$$

If the perimeter of ΔXYZ is three times the perimeter of ΔABC, then its side lengths of ΔXYZ are three times the side lengths of ΔABC. Multiply the vertex matrix of ΔABC by 3. To multiply the matrix by three, multiply each element in the matrix by three.

$$3\begin{bmatrix} -1 & -3 & 0 \\ 4 & -1 & 5 \end{bmatrix} = \begin{bmatrix} -3 & -9 & 0 \\ 12 & -3 & 15 \end{bmatrix}$$

Problem

1. Organize this information into a single matrix: two years ago, while working for the same company, Shaunda received $250, $200, $150, and $350 in quarterly bonuses, while Jerry received $200, $275, $125, and $300.

2. Last year, all employees received twice as much money in quarterly bonuses as they did during previous year. Write and simplify a matrix expression which represents this scenario.

3. Write and simplify a matrix expression showing Shaunda's and Jerry's total bonuses for each quarter of the past two years.

1. A matrix organizes information into rows and columns. Here is one way to organize this data into a 2x4 matrix:

	Q1	Q2	Q3	Q4
Shaunda	$250	$200	$150	$350
Jerry	$200	$275	$125	$300

Another way to organize the information is to use a 4x2 matrix in which the rows represent the quarters and the columns represent the employees.

2. This is an example of scalar multiplication. When multiplying a matrix by a scalar, multiply each term in the matrix by the scale factor, which, in this case, is two.

$$2\begin{bmatrix} \$250 & \$200 & \$150 & \$350 \\ \$200 & \$275 & \$125 & \$300 \end{bmatrix} =$$

$$\begin{bmatrix} 2\cdot\$250 & 2\cdot\$200 & 2\cdot\$150 & 2\cdot\$350 \\ 2\cdot\$200 & 2\cdot\$275 & 2\cdot\$125 & 2\cdot\$300 \end{bmatrix} = \begin{bmatrix} \$500 & \$400 & \$300 & \$700 \\ \$400 & \$550 & \$250 & \$600 \end{bmatrix}$$

3. This is an example of matrix addition. When adding two matrices, the dimensions must be the same. When adding two matrices, add together the corresponding elements of each matrix.

$$\begin{bmatrix} \$250 & \$200 & \$150 & \$350 \\ \$200 & \$275 & \$125 & \$300 \end{bmatrix} + \begin{bmatrix} \$500 & \$400 & \$300 & \$700 \\ \$400 & \$550 & \$250 & \$600 \end{bmatrix} = \begin{bmatrix} \$750 & \$600 & \$450 & \$1050 \\ \$600 & \$825 & \$375 & \$900 \end{bmatrix}$$

2×3 Matrix vs. Square Matrix

1. The dimensions of a matrix are written as row×column, so a 2×3 matrix has two rows and three columns. An example of such a matrix is $\begin{bmatrix} 1 & 2 & 3 \\ 4 & 5 & 6 \end{bmatrix}$.

2. A square matrix has the same number of rows as it has columns. An example of such a matrix is the 3×3 matrix $\begin{bmatrix} 1 & 2 & 3 \\ 4 & 5 & 6 \\ 7 & 8 & 9 \end{bmatrix}$.

Column Matrix and a Row Matrix

A column matrix is a matrix with just one column, such as $\begin{bmatrix} 6 \\ -4 \\ 5 \end{bmatrix}$. A row matrix is a matrix with just one row, such as $\begin{bmatrix} -1 & 8 & 12 & -5 \end{bmatrix}$.

Determine What Must Be True of the Dimensions of Two Matrices in Order for Them to Be Added or Subtracted and Multiplied.

In order for two matrices to be added or subtracted, the matrices must have the same dimensions. The sum or difference has the same dimensions as the matrices which are added or subtracted. In order for two matrices to be multiplied, the number of columns of the first matrix must equal the number of rows of the second matrix. The product has the same number of rows as the first matrix and the same number of columns as the second. For example, a 2×3 matrix can be multiplied by a 3×5 matrix; the resulting product will be a 2×5 matrix.

Problem

Determine whether the following statement is always, sometimes, or never true for matrix [A] with dimensions m×n: $[A]^2$ is defined.

The statement "$[A]^2$ is defined" is sometimes true: the square of a matrix is only defined for square matrices, or matrices where $m = n$. In order for two matrices to be multiplied, the number of rows of the second matrix must equal the number of columns of the first.

Problem

Determine whether or not the commutative property of addition is true for square matrices. If true, provide an explanation; if false. provide a counterexample.

Matrix addition is commutative. Consider the 2x2 matrices $[A] = \begin{bmatrix} a_1 & a_2 \\ a_3 & a_4 \end{bmatrix}$ and $[B] = \begin{bmatrix} b_1 & b_2 \\ b_3 & b_4 \end{bmatrix}$. Because these two matrices have the same dimensions, they can be added by adding corresponding elements together: $[A] + [B] = \begin{bmatrix} a_1 + b_1 & a_2 + b_2 \\ a_3 + b_3 & a_4 + b_4 \end{bmatrix}$, and $[B] + [A] = \begin{bmatrix} b_1 + a_1 & b_2 + a_2 \\ b_3 + a_3 & b_4 + a_4 \end{bmatrix}$. The commutative property of addition states that $a_1 + b_1 = b_1 + a_1$; $a_2 + b_2 = b_2 + a_2$; and so on.

Thus, $\begin{bmatrix} a_1 + b_1 & a_2 + b_2 \\ a_3 + b_3 & a_4 + b_4 \end{bmatrix} = \begin{bmatrix} b_1 + a_1 & b_2 + a_2 \\ b_3 + a_3 & b_4 + a_4 \end{bmatrix}$, so $[A] + [B] = [B] + [A]$. Matrix addition is commutative not only for 2×2 square matrices [A] and [B] but also for all matrices [A] and [B] whose dimensions are the same.

Problem

Determine whether or not the commutative property of multiplication is true for square matrices. If true, provide an explanation; if false, provide a counterexample.

Matrix multiplication is NOT commutative. Consider the 2x2 matrices $[A] = \begin{bmatrix} 1 & 2 \\ 3 & 4 \end{bmatrix}$ and $[B] = \begin{bmatrix} 5 & 6 \\ 7 & 8 \end{bmatrix}$. Because the number of columns of the first matrix is the same as the number of rows of the second, the two matrices can be multiplied. Using the rules for matrix multiplication, $[A] \cdot [B] = \begin{bmatrix} 1 & 2 \\ 3 & 4 \end{bmatrix} \cdot \begin{bmatrix} 5 & 6 \\ 7 & 8 \end{bmatrix} =$
$\begin{bmatrix} 1 \cdot 5 + 2 \cdot 7 & 1 \cdot 6 + 2 \cdot 8 \\ 3 \cdot 5 + 4 \cdot 7 & 3 \cdot 6 + 4 \cdot 8 \end{bmatrix} = \begin{bmatrix} 19 & 22 \\ 43 & 50 \end{bmatrix}$, while $[B] \cdot [A] = \begin{bmatrix} 5 & 6 \\ 7 & 8 \end{bmatrix} \cdot \begin{bmatrix} 1 & 2 \\ 3 & 4 \end{bmatrix}$

$= \begin{bmatrix} 5 \cdot 1 + 6 \cdot 3 & 5 \cdot 2 + 6 \cdot 4 \\ 7 \cdot 1 + 8 \cdot 3 & 7 \cdot 2 + 8 \cdot 4 \end{bmatrix} = \begin{bmatrix} 23 & 34 \\ 31 & 46 \end{bmatrix}$. The resulting products are not the same.Therefore, $[A] \cdot [B] \neq [B] \cdot [A]$.

Problem

Determine whether or not the associative property of addition is true for square matrices. If true, provide an explanation; if false, provide a counterexample.

The associative property of addition is true for matrices. Consider the 2x2 matrices $[A] = \begin{bmatrix} a_1 & a_2 \\ a_3 & a_4 \end{bmatrix}$ and $[B] = \begin{bmatrix} b_1 & b_2 \\ b_3 & b_4 \end{bmatrix}$ and $[C] = \begin{bmatrix} c_1 & c_2 \\ c_3 & c_4 \end{bmatrix}$.

$$[A] + ([B] + [C]) = \begin{bmatrix} a_1 & a_2 \\ a_3 & a_4 \end{bmatrix} + (\begin{bmatrix} b_1 & b_2 \\ b_3 & b_4 \end{bmatrix} + \begin{bmatrix} c_1 & c_2 \\ c_3 & c_4 \end{bmatrix})$$

$$= \begin{bmatrix} a_1 & a_2 \\ a_3 & a_4 \end{bmatrix} + \begin{bmatrix} b_1 + c_1 & b_2 + c_2 \\ b_3 + c_3 & b_4 + c_4 \end{bmatrix} = \begin{bmatrix} a_1 + b_1 + c_1 & a_2 + b_2 + c_2 \\ a_3 + b_3 + c_3 & a_4 + b_4 + c_4 \end{bmatrix}$$

$([A] + [B]) + [C] = (\begin{bmatrix} a_1 & a_2 \\ a_3 & a_4 \end{bmatrix} + \begin{bmatrix} b_1 & b_2 \\ b_3 & b_4 \end{bmatrix}) + \begin{bmatrix} c_1 & c_2 \\ c_3 & c_4 \end{bmatrix} = \begin{bmatrix} a_1 + b_1 & a_2 + b_2 \\ a_3 + b_3 & a_4 + b_4 \end{bmatrix} + \begin{bmatrix} c_1 & c_2 \\ c_3 & c_4 \end{bmatrix} = \begin{bmatrix} a_1 + b_1 + c_1 & a_2 + b_2 + c_2 \\ a_3 + b_3 + c_3 & a_4 + b_4 + c_4 \end{bmatrix}$.

The two resulting matrices are the same, so $[A] + ([B] + [C]) = ([A] + [B]) + [C]$. Matrix addition is associative not only for 2×2 square matrices [A], [B], and [C] but also for all matrices [A], [B], and [C] whose dimensions are the same.

Problem

Determine whether or not the associative property of multiplication is true for square matrices. If true, provide an explanation; if false, provide a counterexample.

The associative property of multiplication is true for matrices. Consider the 2x2 matrices $[A] = \begin{bmatrix} a_1 & a_2 \\ a_3 & a_4 \end{bmatrix}$ and $[B] = \begin{bmatrix} b_1 & b_2 \\ b_3 & b_4 \end{bmatrix}$ and $[C] = \begin{bmatrix} c_1 & c_2 \\ c_3 & c_4 \end{bmatrix}$.

$$[A] \cdot ([B] \cdot [C]) = \begin{bmatrix} a_1 & a_2 \\ a_3 & a_4 \end{bmatrix} \cdot (\begin{bmatrix} b_1 & b_2 \\ b_3 & b_4 \end{bmatrix} \cdot \begin{bmatrix} c_1 & c_2 \\ c_3 & c_4 \end{bmatrix})$$

$$= \begin{bmatrix} a_1 & a_2 \\ a_3 & a_4 \end{bmatrix} \cdot \begin{bmatrix} b_1c_1 + b_2c_3 & b_1c_2 + b_2c_4 \\ b_3c_1 + b_4c_3 & b_3c_2 + b_4c_4 \end{bmatrix}$$

$$= \begin{bmatrix} a_1(b_1c_1 + b_2c_3) + a_2(b_3c_1 + b_4c_3) & a_1(b_1c_2 + b_2c_4) + a_2(b_3c_2 + b_4c_4) \\ a_3(b_1c_1 + b_2c_3) + a_4(b_3c_1 + b_4c_3) & a_3(b_1c_2 + b_2c_4) + a_4(b_3c_2 + b_4c_4) \end{bmatrix}$$

$$= \begin{bmatrix} a_1b_1c_1 + a_1b_2c_3 + a_2b_3c_1 + a_2b_4c_3 & a_1b_1c_2 + a_1b_2c_4 + a_2b_3c_2 + a_2b_4c_4 \\ a_3b_1c_1 + a_3b_2c_3 + a_4b_3c_1 + a_4b_4c_3 & a_3b_1c_2 + a_3b_2c_4 + a_4b_3c_2 + a_4b_4c_4 \end{bmatrix}.$$

$$([A] \cdot [B]) \cdot [C] = \left(\begin{bmatrix} a_1 & a_2 \\ a_3 & a_4 \end{bmatrix} \cdot \begin{bmatrix} b_1 & b_2 \\ b_3 & b_4 \end{bmatrix}\right) \cdot \begin{bmatrix} c_1 & c_2 \\ c_3 & c_4 \end{bmatrix}$$

$$= \begin{bmatrix} a_1b_1 + a_2b_3 & a_1b_2 + a_2b_4 \\ a_3b_1 + a_4b_3 & a_3b_2 + a_4b_4 \end{bmatrix} \cdot \begin{bmatrix} c_1 & c_2 \\ c_3 & c_4 \end{bmatrix}$$

$$= \begin{bmatrix} (a_1b_1 + a_2b_3)c_1 + (a_1b_2 + a_2b_4)c_3 & (a_1b_1 + a_2b_3)c_2 + (a_1b_2 + a_2b_4)c_4 \\ (a_3b_1 + a_4b_3)c_1 + (a_3b_2 + a_4b_4)c_3 & (a_3b_1 + a_4b_3)c_2 + (a_3b_2 + a_4b_4)c_4 \end{bmatrix}$$

$$= \begin{bmatrix} a_1b_1c_1 + a_2b_3c_1 + a_1b_2c_3 + a_2b_4c_3 & a_1b_1c_2 + a_2b_3c_2 + a_1b_2c_4 + a_2b_4c_4 \\ a_3b_1c_1 + a_4b_3c_1 + a_3b_2c_3 + a_4b_4c_3 & a_3b_1c_2 + a_4b_3c_2 + a_3b_2c_4 + a_4b_4c_4 \end{bmatrix}.$$

The two resulting matrices are equivalent, so $[A] \cdot ([B] \cdot [C]) = ([A] \cdot [B]) \cdot [C]$. Matrix multiplication is associative not only for 2×2 square matrices [A],

[B], and[C]but also for all matrices [A], [B], and[C]such that the number of rows of [B] equals the number of columns of [A] and the number of columns of [B] equals the number of rows of [C]; the resulting product will contain the number of rows of [A] and the number of columns of [C].

ZERO MATRIX VS. IDENTITY MATRIX

1. A zero matrix, also called a null matrix, is a matrix in which all entries are zero; it can be added to another matrix having the same dimensions without changing the other matrix's identity. An example of a zero matrix is $\begin{bmatrix} 0 & 0 \\ 0 & 0 \\ 0 & 0 \end{bmatrix}$, which is the additive identity for 3×2 matrices.

2. An identity matrix, also called a unit matrix, is a square matrix in which all of the entries along the main diagonal are ones and all other entries in the matrix are zeros. An identity matrix can be multiplied by another matrix of appropriate dimensions without changing the other matrix's identity, so it is a multiplicative identity. An example of an identity matrix is $\begin{bmatrix} 1 & 0 & 0 \\ 0 & 1 & 0 \\ 0 & 0 & 1 \end{bmatrix}$.

Problem

Determine whether or not $[A] \cdot ([B] + [C]) = [A] \cdot [B] + [A] \cdot [C]$ is true for square matrices [A], [B], and [C]. If true, provide an explanation; if false, provide a counterexample.

The distributive property is true for square matrices. Consider the 2x2 matrices $[A] = \begin{bmatrix} a_1 & a_2 \\ a_3 & a_4 \end{bmatrix}$ and $[B] = \begin{bmatrix} b_1 & b_2 \\ b_3 & b_4 \end{bmatrix}$ and $[C] = \begin{bmatrix} c_1 & c_2 \\ c_3 & c_4 \end{bmatrix}$.

$$[A] \cdot ([B] + [C]) = \begin{bmatrix} a_1 & a_2 \\ a_3 & a_4 \end{bmatrix}\left(\begin{bmatrix} b_1 & b_2 \\ b_3 & b_4 \end{bmatrix} + \begin{bmatrix} c_1 & c_2 \\ c_3 & c_4 \end{bmatrix}\right) = \begin{bmatrix} a_1 & a_2 \\ a_3 & a_4 \end{bmatrix}\left(\begin{bmatrix} b_1 + c_1 & b_2 + c_2 \\ b_3 + c_3 & b_4 + c_4 \end{bmatrix}\right)$$

$$= \begin{bmatrix} a_1(b_1 + c_1) + a_2(b_3 + c_3) & a_1(b_2 + c_2) + a_2(b_4 + c_4) \\ a_3(b_1 + c_1) + a_4(b_3 + c_3) & a_3(b_2 + c_2) + a_4(b_4 + c_4) \end{bmatrix}$$

$$= \begin{bmatrix} a_1b_1 + a_1c_1 + a_2b_3 + a_2c_3 & a_1b_2 + a_1c_2 + a_2b_4 + a_2c_4 \\ a_3b_1 + a_3c_1 + a_4b_3 + a_4c_3 & a_3b_2 + a_3c_2 + a_4b_4 + a_4c_4 \end{bmatrix}$$

$$[A] \cdot [B] + [A] \cdot [C] = \begin{bmatrix} a_1 & a_2 \\ a_3 & a_4 \end{bmatrix} \cdot \begin{bmatrix} b_1 & b_2 \\ b_3 & a_2 \end{bmatrix} + \begin{bmatrix} a_1 & a_2 \\ a_3 & a_4 \end{bmatrix} \cdot \begin{bmatrix} c_1 & c_2 \\ c_3 & c_4 \end{bmatrix}$$

$$= \begin{bmatrix} a_1b_1 + a_2b_3 & a_1b_2 + a_2b_4 \\ a_3b_1 + a_4b_3 & a_3b_2 + a_4b_4 \end{bmatrix} + \begin{bmatrix} a_1c_1 + a_2c_3 & a_1c_2 + a_2c_4 \\ a_3c_1 + a_4c_3 & a_3c_2 + a_4c_4 \end{bmatrix}$$

$$= \begin{bmatrix} a_1b_1 + a_2b_3 + a_1c_1 + a_2c_3 & a_1b_2 + a_2b_4 + a_1c_2 + a_2c_4 \\ a_3b_1 + a_4b_3 + a_3c_1 + a_4c_3 & a_3b_2 + a_4b_4 + a_3c_2 + a_4c_4 \end{bmatrix}$$

The two resulting matrices are equivalent, so $[A] \cdot ([B] + [C]) = [A] \cdot [B] + [A] \cdot [C]$. This statement is true not only for 2×2 square matrices [A], [B], and[C] but also for all matrices [A], [B], and [C] such that the dimensions of [B] and [C] are the same and such that the number of columns of [A] equals the number of rows of [B] and [C].

Problem

Write the additive inverse of matrix $\begin{bmatrix} 2 & 3 & 0 \\ -1 & 4 & -8 \end{bmatrix}$.

The sum of a matrix and its additive inverse is the zero matrix. In order for two matrices to be added, their dimensions must be equal. The resulting sum will have the same dimensions as the two added matrices. The additive inverse of a 2×3 is the 2×3 matrix which, when added to the first, gives the 2×3 matrix $\begin{bmatrix} 0 & 0 & 0 \\ 0 & 0 & 0 \end{bmatrix}$. The corresponding elements in a matrix and in its additive inverse matrix are additive inverses. So, the additive inverse of the given matrix is $\begin{bmatrix} -2 & -3 & 0 \\ 1 & -4 & 8 \end{bmatrix}$.

$$\begin{bmatrix} 2 & 3 & 0 \\ -1 & 4 & -8 \end{bmatrix} + \begin{bmatrix} -2 & -3 & 0 \\ 1 & -4 & 8 \end{bmatrix} = \begin{bmatrix} 0 & 0 & 0 \\ 0 & 0 & 0 \end{bmatrix}$$

INVERSE AND DETERMINANT IN TERMS OF MATRICES

The terms *inverse* and *determinant* relate only to square matrices, or matrices which have the same number of rows and columns. A matrix is invertible only if its determinant is not zero. The product of a matrix and its inverse is equal to the identity matrix, a square matrix in which all of the entries along the main diagonal are ones and all other entries in the matrix are zeros.

Example

Find the determinant of a 2×2 matrix and of a 3×3 matrix.

Consider 2×2 matrix $\begin{bmatrix} a & b \\ c & d \end{bmatrix}$. The determinant is written $\begin{vmatrix} a & b \\ c & d \end{vmatrix}$ and has the value $\boldsymbol{ad - bc}$.

Consider the 3×3 matrix $\begin{bmatrix} a & b & c \\ d & e & f \\ g & h & i \end{bmatrix}$.The determinant is written $\begin{vmatrix} a & b & c \\ d & e & f \\ g & h & i \end{vmatrix}$ and has the value $a\begin{vmatrix} e & f \\ h & i \end{vmatrix} - b\begin{vmatrix} d & f \\ g & i \end{vmatrix} + c\begin{vmatrix} d & e \\ g & h \end{vmatrix} = a(ei - fh) - b(di - fg) + c(dh - eg) = aei - afh - bdi + bfg + cdh - ceg$Notice that this is the sum of the products of the terms of the three top/left-to-bottom/right diagonals minus the sum of the products of the terms of the three top/right-to-bottom/left diagonals:$(\boldsymbol{aei + bfg + cdh}) - (\boldsymbol{ceg + bdi + afh})$.

Example

Example of a square matrix which has an inverse and a square matrix which has no inverse.

A square matrix has an inverse if the value of its determinant is not zero and has no inverse if the value of its determinant is zero. For example, the matrix $\begin{bmatrix} 1 & 2 \\ 3 & 4 \end{bmatrix}$ has an inverse because its determinant is $\begin{vmatrix} 1 & 2 \\ 3 & 4 \end{vmatrix} = 1 \cdot 4 - 2 \cdot 3 = 4 - 6 = -2$. The matrix $\begin{bmatrix} 3 & 2 \\ 6 & 4 \end{bmatrix}$ has no inverse because its determinant is $\begin{vmatrix} 3 & 2 \\ 6 & 4 \end{vmatrix} = 3 \cdot 4 - 2 \cdot 6 = 12 - 12 = 0$.

Problem

Simplify $\begin{bmatrix} 2 & 0 & -5 \end{bmatrix}\begin{bmatrix} 1 \\ -3 \\ 4 \end{bmatrix}$.

The product of a 1×3 matrix a 3×1 matrix is a 1×1 matrix.

$$\begin{bmatrix} 2 & 0 & -5 \end{bmatrix}\begin{bmatrix} 1 \\ -3 \\ 4 \end{bmatrix} = [(2)(1) + (0)(-3) + (-5)(4)] = [-18]$$

Problem

Simplify $\begin{bmatrix} 1 \\ -3 \\ 4 \end{bmatrix}\begin{bmatrix} 2 & 0 & -5 \end{bmatrix}$.

The product of a 3×1 matrix a 1×3 matrix is a 3×3 matrix.

$$\begin{bmatrix} 1 \\ -3 \\ 4 \end{bmatrix}\begin{bmatrix} 2 & 0 & -5 \end{bmatrix} = \begin{bmatrix} (1)(2) & (1)(0) & (1)(-5) \\ (-3)(2) & (-3)(0) & (-3)(-5) \\ (4)(2) & (4)(0) & (4)(-5) \end{bmatrix} = \begin{bmatrix} 2 & 0 & -5 \\ -6 & 0 & 15 \\ 8 & 0 & -20 \end{bmatrix}$$

Problem

Simplify $\begin{bmatrix} 2 & 1 & -5 \\ -6 & 3 & 15 \\ 8 & 6 & -20 \end{bmatrix}\left(\begin{bmatrix} 2 & 1 & -5 \\ -6 & 3 & 15 \\ 8 & 6 & -20 \end{bmatrix} - \begin{bmatrix} 1 & 1 & -5 \\ -6 & 2 & 15 \\ 8 & 6 & -21 \end{bmatrix}\right)$

$$\begin{bmatrix} 2 & 1 & -5 \\ -6 & 3 & 15 \\ 8 & 6 & -20 \end{bmatrix}\left(\begin{bmatrix} 2 & 1 & -5 \\ -6 & 3 & 15 \\ 8 & 6 & -20 \end{bmatrix} - \begin{bmatrix} 1 & 1 & -5 \\ -6 & 2 & 15 \\ 8 & 6 & -21 \end{bmatrix}\right)$$

$$= \begin{bmatrix} 2 & 1 & -5 \\ -6 & 3 & 15 \\ 8 & 6 & -20 \end{bmatrix}\left(\begin{bmatrix} 2-1 & 1-1 & -5-(-5) \\ -6-(-6) & 3-2 & 15-15 \\ 8-8 & 6-6 & -20-(-21) \end{bmatrix}\right)$$

$$= \begin{bmatrix} 2 & 1 & -5 \\ -6 & 3 & 15 \\ 8 & 6 & -20 \end{bmatrix}\begin{bmatrix} 1 & 0 & 0 \\ 0 & 1 & 0 \\ 0 & 0 & 1 \end{bmatrix} = \begin{bmatrix} 2 & 1 & -5 \\ -6 & 3 & 15 \\ 8 & 6 & -20 \end{bmatrix}$$

Problem

Represent the vector as a matrix.

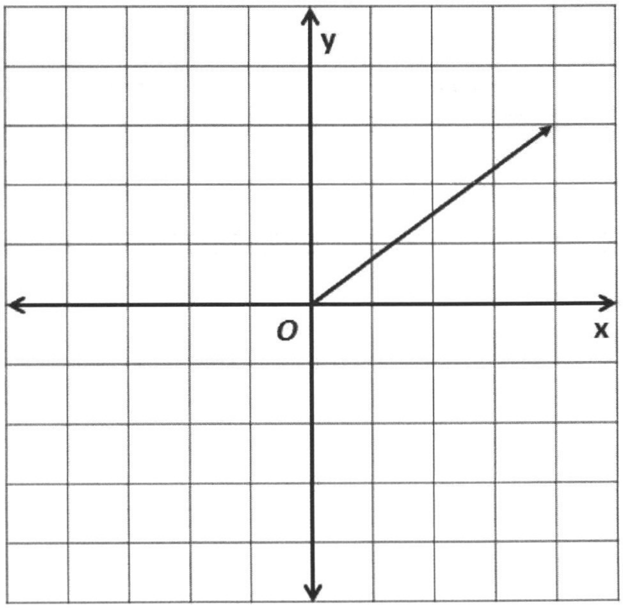

A vector can be represented as a column matrix whose elements are the magnitudes of its component vectors. Thus, this vector can be represented by the matrix $\begin{bmatrix} 4 \\ 3 \end{bmatrix}$.

Example

1. Consider the transformation $\begin{bmatrix} x \\ y \end{bmatrix} \rightarrow \begin{bmatrix} x' \\ y' \end{bmatrix}$**. If** $x' = ax + by$ **and** $y' = cx + dy$**, explain how matrix** $\begin{bmatrix} a & b \\ c & d \end{bmatrix}$ **can be used to transform** $\begin{bmatrix} x \\ y \end{bmatrix}$ **into** $\begin{bmatrix} x' \\ y' \end{bmatrix}$.

2. If $\begin{bmatrix} a & b \\ c & d \end{bmatrix} = \begin{bmatrix} 2 & 3 \\ 1 & -2 \end{bmatrix}$ **and** $\begin{bmatrix} x \\ y \end{bmatrix} = \begin{bmatrix} 3 \\ 4 \end{bmatrix}$**, find** $\begin{bmatrix} x' \\ y' \end{bmatrix}$.

1. Multiply the 2×2 transformation matrix $\begin{bmatrix} a & b \\ c & d \end{bmatrix}$ **by the original 2×1 position matrix** $\begin{bmatrix} x \\ y \end{bmatrix}$ **to find the 2x1 matrix** $\begin{bmatrix} x' \\ y' \end{bmatrix}$**:**

$$\begin{bmatrix} a & b \\ c & d \end{bmatrix} \begin{bmatrix} x \\ y \end{bmatrix} = \begin{bmatrix} ax + by \\ cx + dy \end{bmatrix}$$

Since $ax + by = x'$, and $cx + dy = y'$, $\begin{bmatrix} a & b \\ c & d \end{bmatrix} \begin{bmatrix} x \\ y \end{bmatrix} = \begin{bmatrix} x' \\ y' \end{bmatrix}$.

2. $\begin{bmatrix} x' \\ y' \end{bmatrix} = \begin{bmatrix} 2 & 3 \\ 1 & -2 \end{bmatrix} \begin{bmatrix} 3 \\ 4 \end{bmatrix}$

$$\begin{bmatrix} x' \\ y' \end{bmatrix} = \begin{bmatrix} 2(3) + (3)(4) \\ 1(3) + (-2)(4) \end{bmatrix} = \begin{bmatrix} 6 + 12 \\ 3 - 8 \end{bmatrix} = \begin{bmatrix} 18 \\ -5 \end{bmatrix}$$

Transformation Matrix

Transformation Matrix Which Reflects a Vector Across the X-Axis

To write a matrix which transforms a vector, first write the equations $x' = ax + by$ and $y' = cx + dy$. Then, organize $a, b, c,$and d into transformation matrix and organize into transformation matrix $\begin{bmatrix} a & b \\ c & d \end{bmatrix}$.

Consider point (x, y) and its reflection across the x-axis (x', y'). The x-value of the reflected point is the same as the x-value of the original point, so $x' = x$, which is the same as $x' = 1x + 0y$. The y-value of the reflected point is the opposite of the y-value of the original point, so $y' = -y$, which is the same as $y' = 0x - 1y$. So, $\begin{bmatrix} 1 & 0 \\ 0 & -1 \end{bmatrix}$ is the transformation matrix which reflects a vector across the x-axis.

Transformation Matrix Which Rotates a Vector 180° About the Origin

To write a matrix which transforms a vector, first write the equations $x' = ax + by$ and$y' = cx + dy$. Then, organize $a, b, c,$and d into transformation matrix and organize into transformation matrix $\begin{bmatrix} a & b \\ c & d \end{bmatrix}$.

Consider point (x, y) and (x', y'), its rotation about the origin. The x-value of the rotated point is the opposite as the x-value of the original point, so $x' = -x$, which is the same as $x' = -1x + 0y$. The y-value of the rotated point is also the opposite of the y-value of the original point, so $y' = -y$, which is the same as $y' = 0x - 1y$. So, $\begin{bmatrix} -1 & 0 \\ 0 & -1 \end{bmatrix}$ is the transformation matrix which rotates a vector about the origin.

Example

Find the number of seconds in a week.

To find the number of seconds in a week, start with the number of days in a week and use the units as a guide to solve the problem. This problem solving strategy is called dimensional analysis.

$$\frac{7\ \cancel{\text{days}}}{\textbf{week}} \cdot \frac{24\ \cancel{\text{hours}}}{\cancel{\text{day}}} \cdot \frac{60\ \cancel{\text{minutes}}}{\cancel{\text{hour}}} \cdot \frac{60\ \textbf{seconds}}{\cancel{\text{minute}}} = \frac{604{,}800\ \text{seconds}}{\text{week}}$$

Example

Find the mass in grams of a sample which contains 1.807×10^{24} molecules of sodium chloride (NaCl). The mass of one mole of NaCl is approximately 58.4 g, and there are 6.022×10^{23} molecules in one mole of NaCl.

Start with the given quantity (molecules of NaCl) and use units to find the unknown quantity (g of NaCl):

$$\frac{1.807 \times 10^{24}\ \cancel{\text{molecules NaCl}}}{1} \cdot \frac{1\ \cancel{\text{mol NaCl}}}{6.022 \times 10^{23}\ \cancel{\text{molecules NaCl}}} \cdot \frac{58.4\ \text{g NaCl}}{1\ \cancel{\text{mol NaCl}}} = 175\text{g NaCl}$$

There are 175 grams of sodium chloride in a sample which contains 1.807×10^{24} molecules.

FINDING THE AREA OF PARALLELOGRAM *ABCD* USING A DETERMINANT

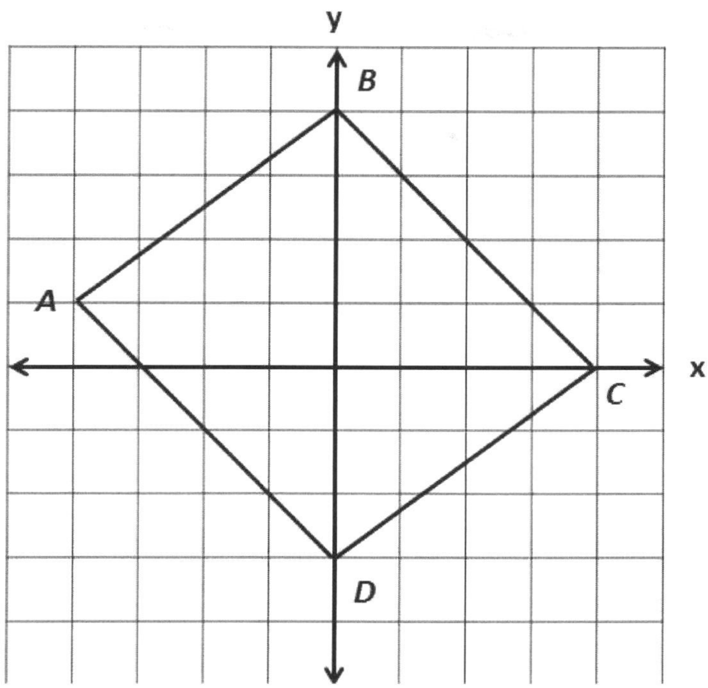

Choose any two sides of the parallelogram which share a vertex. For instance, sides $\overline{AB}$ and $\overline{AD}$ share a vertex at (-4,1).

Find the components of vectors $\overrightarrow{AB}$ and $\overrightarrow{AD}$ (or whichever sides were chosen in step 1). The components of a vector can be found by subtracting its initial point from its terminal point:

$$\overrightarrow{AB} = (0,4) - (-4,1) = (0+4, 4-1) = (4,3)$$

$$\overrightarrow{AD} = (0,-3) - (-4,1) = (0+4,-3-1) = (4,-4)$$

Organize the components into a 2x2 matrix.

$$\begin{bmatrix} 4 & 3 \\ 4 & -4 \end{bmatrix}$$

Find the area of the parallelogram by finding the absolute value of the determinant of the matrix.

$$\text{abs}\begin{vmatrix} 4 & 3 \\ 4 & -4 \end{vmatrix} = \text{abs}(4 \cdot -4 - 3 \cdot 4) = \text{abs}(-16 - 12) = 28$$

The area of the parallelogram is 28 square units.

Example

Create a graph which models the population growth of a bacterial culture over a three hour time period. Suppose that there are presently 100 bacteria in the culture and that the doubling rate is approximately thirty minutes.

The number of bacteria in the culture depends on how much time has passed. Therefore, the time in minutes is plotted along the x-axis while the number of bacteria is plotted along the y-axis.

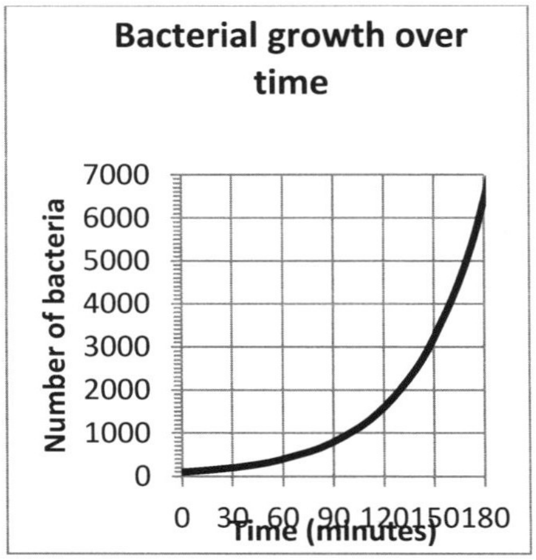

Problem

A girl walks at an average pace of three miles per hour. Draw a graph which represents the distance she travels over time.

The distance traveled depends on the travel time, so distance is the dependent variable and is graphed on the vertical axis, while time is the independent variable and is graphed along the horizontal axis. The linear equation $d = rt$, where d is distance traveled, r is rate in mph, and t is time in hours, can be used to represent this scenario. The graph, however, is only a first quadrant graph since neither travel time nor distance traveled can be negative.

Problem

Determine a quantity which can be used to report

1. Highway safety

2. Efficacy of a pain reliever.

1. One measure of highway safety is the number of fatalities per year on that particular highway.

2. One measure of a pain reliever's efficacy is the percentage of patients who report pain relief after taking the drug.

Problem

1. Report the volume of a liquid measured in a graduated cylinder with 100 mL demarcations

2. Report the length of an object measured using a ruler with 1 cm demarcations.

The last digit of a reported quantity is estimated and is therefore uncertain.

1. Report the measurement of the volume of a liquid measured in a graduated cylinder with 100 mL demarcations to the nearest 10 mL. The volume is measured from the bottom of the meniscus.

2. Report the measurement of the length of an object measured using a ruler with 1 cm demarcations to the nearest mm.

Practice Test #1

Practice Questions

1. Simplify the expression $\left(5^{1/3} \cdot 5^2\right)^6$.

a. $5^{2/3}$
b. 5^4
c. $5^{\left(8\frac{1}{3}\right)}$
d. 5^{14}

2. Which of the following is false?

a. $\left(8^{1/2}\right)^2 = 2^3$
b. $4 \cdot 4 \cdot 4 \cdot 4 \cdot \sqrt{4} = 4^{9/2}$
c. $5^{1/3} \cdot 5^{2/3} \cdot 5^{1/3} = 5\sqrt{5}$
d. $\frac{\sqrt[6]{5}}{\sqrt[6]{5^2}} = 5^{-1/6}$

3. Rewrite using fractional exponents: $5\sqrt[3]{2^2} + \sqrt[7]{3^2}$.

a. $5 \cdot 2^{2/3} + 3$
b. $5 \cdot 2^{2/3} + 3^{2/7}$
c. $5^3 \cdot 2^{1/2} + 7 \cdot 3^{1/2}$
d. $10^{2/3} + 3^{2/7}$

4. Simplify: $\frac{5^8 \cdot \sqrt[3]{5^2} \cdot 5^{-3}}{\sqrt[3]{5^5}}$.

a. $5^5 \cdot 5^{7/3}$
b. 5^6
c. 5^4
d. $5^{2/3}$

5. Which of the following is a rational number?

a. $3\sqrt{121} + 7$
b. $4 + \sqrt[3]{150}$
c. $7\sqrt{30}$
d. $(8)^{2/3} + (8)^{1/5}$

6. Derek is purchasing lunch meat for the class picnic. There are 300 people attending the picnic. Derek expects that for every 10 people 11 sandwiches will be eaten and that 20 pounds of meat will make 90 sandwiches. How much lunch meat should Derek buy?

a. 73 ⅓ pounds
b. 67 ⅔ pounds
c. 110 pounds
d. 330 sandwiches

7. A certain radioactive material has a half-life of 7 days. A test sample of this material has an initial mass of 400 kg. Which of the following would be the most appropriate unit to use when labeling the time axis of a graph of remaining material vs. time elapsed until 6.1 grams remain?

a. Hours
b. Days
c. Weeks
d. Years

8. Usain Bolt set a world record of 9.58 seconds for the 100 meter dash. White-tailed deer can run at 48 kilometers per hour for short periods. If Mr. Bolt and a white-tailed deer run a 100 meter dash at these speeds, who would win, and by how much?

a. Mr. Bolt wins by 3.75 seconds
b. Mr. Bolt wins by 2.08 seconds
c. Deer wins by 3.75 seconds
d. Deer wins by 2.08 seconds

9. After getting his wisdom teeth pulled, Matt was prescribed 60 mg of Codeine to be taken every 4 hours for 5 days. The pharmacy has 30 mg pills. How many pills should Matt receive when filling the prescription?

a. 72
b. 60
c. 30
d. 24

10. Label the axes and choose a scale for the graphic model of this problem:

Meghan bought cookies and milk for her classmates. She spent \$.45 per cookie, one cookie each per student, and \$2.65 on one gallon of milk to share. How much does she spend altogether?

a.
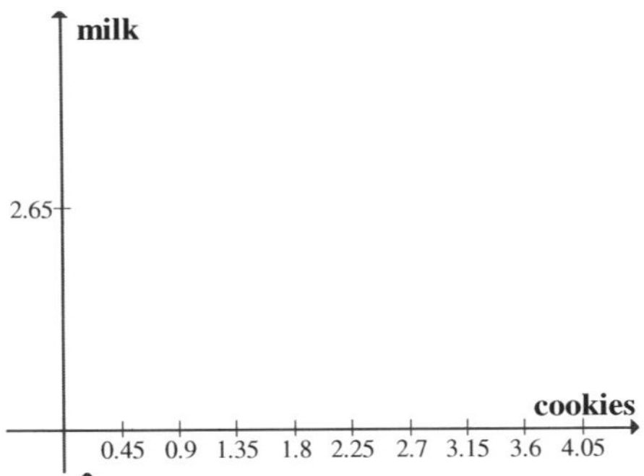

b.
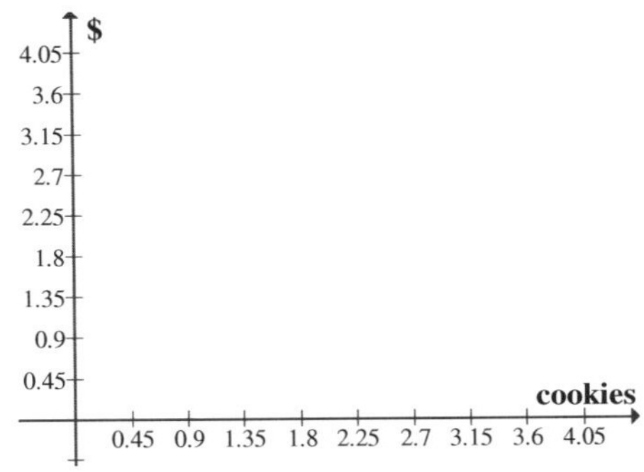

c.
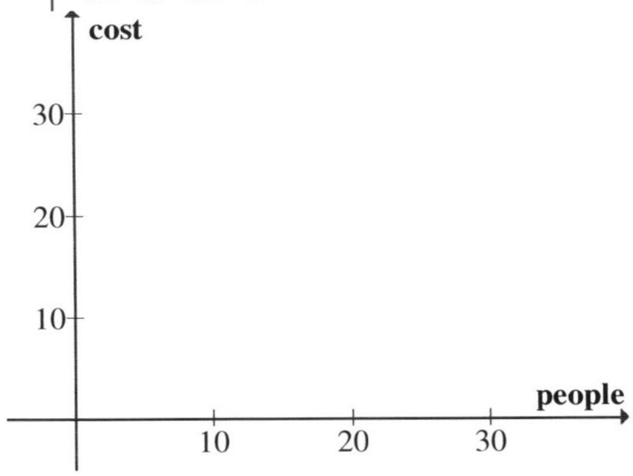

d.
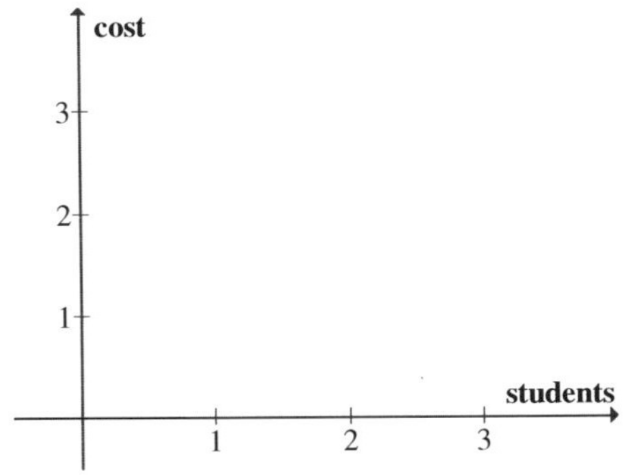

11. Simplify: $3 + \sqrt{3^2 - 4 \cdot 5}$.

a. $3 + \sqrt{11}$
b. $6 - \sqrt{20}$
c. $3 + i\sqrt{11}$
d. $6 - i\sqrt{11}$

12. Simplify: $(3 - 6i)(5 + 4i)$

a. $39 - 18i$
b. $15 - 24i$
c. $-9 - 18i$
d. $-18i^2$

13. Simplify: $3(6 + 2i) - 4(1 + i)$

a. $14 + 3i$
b. $14 + 2i$
c. $22 + 10i$
d. $13 + i$

14. Solve: $\frac{(3+4i)}{(5-2i)}$

a. $\frac{3}{5} - 2i$
b. $\frac{7+26i}{29}$
c. $\frac{17+26i}{25}$
d. $\frac{23+26i}{21}$

15. Compute the modulus of $6 + 8i$.

a. -10
b. -28
c. 10
d. 28

16. Which of the following is the correct graphical representation of $3 + 2i$?

a.
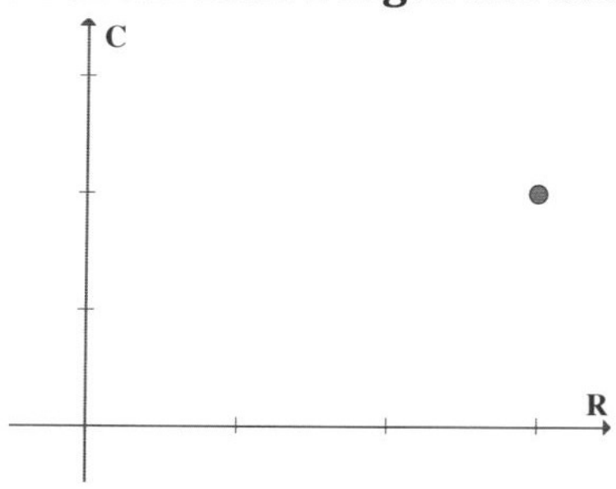

b.
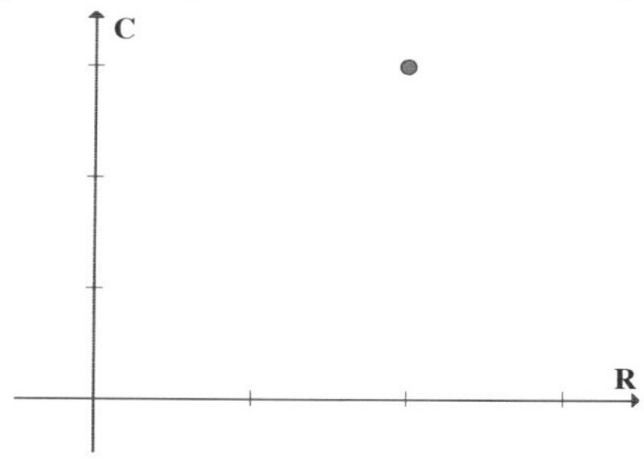

c.
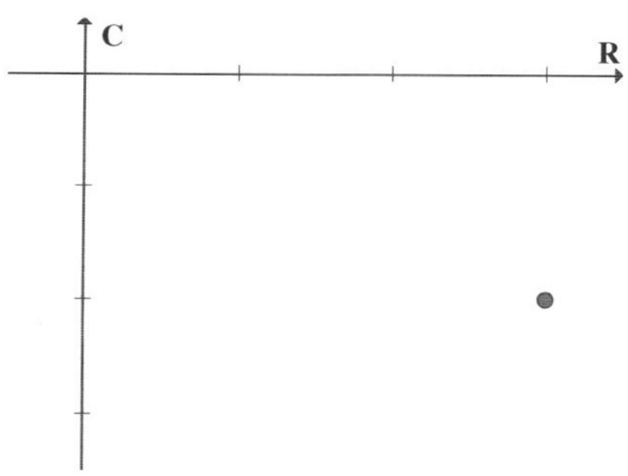

d.
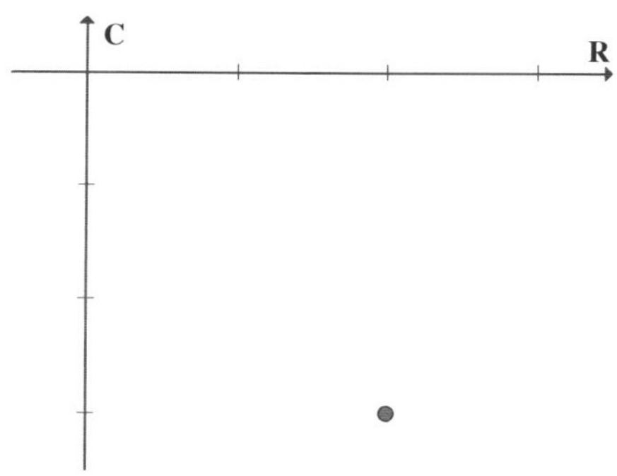

17. Convert from rectangular to polar form: $z = 12 + 5i$.

a. $z = 12(cos5° + isin5°)$
b. $z = 13(cos67° + isin67°)$
c. $z = 17(cos60° + isin60°)$
d. $z = 13(cos23° + isin23°)$

18. Complex numbers are sometimes drawn as vectors to allow for addition and subtraction on the graph. Identify the equation of complex numbers represented by the following graph of vectors:

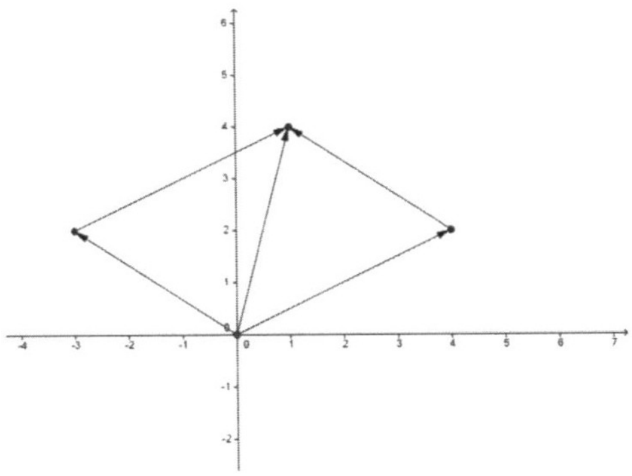

a. $(-3 + 2i) + (4 + 2i) = 1 + 4i$
b. $(4 + 2i) - (-3 + 2i) = 1 + 4i$
c. $(1 + 4i) - (-3 + 2i) = 4 + 2i$
d. $(2 - 3i) + (2 + 4i) = 4 + i$

19. Multiply $(12 + 5i)(-4 + 3i)$. Express the answer in polar form.

a. $18(\cos(-11°) + i\sin(-11°)$
b. $65(\cos(-11°) + i\sin(-11°))$
c. $18(\cos 166° + i\sin 166°)$
d. $65(\cos 166° + i\sin 166°)$

20. Find the midpoint between $-30 + 15i$ and $12 - 3i$.

a. $9 + 6i$
b. $-9 + 6i$
c. $-21 + 9i$
d. $-24 + 13.5i$

21. Find the distance, rounded to the nearest integer, between $-30 + 15i$ and $12 - 3i$.

a. 22
b. 44
c. 46
d. 51

22. Solve for all roots: $2x^2 = 5x - 11$.

a. $\frac{-5 \pm i\sqrt{113}}{4}$
b. $\frac{5 \pm \sqrt{113}}{4}$
c. $\frac{5 \pm 3i\sqrt{7}}{4}$
d. $\frac{-5 \pm 3\sqrt{7}}{4}$

23. Find the complex roots: $y = x^2 - 2x + 10$.

a. $1 + 3i, 1 - 3i$
b. $1 + 6i, 1 - 6i$
c. $1 + i\sqrt{22}, 1 - i\sqrt{22}$
d. $2 + i\sqrt{22}, 2 - i\sqrt{22}$

24. Factor: $4x^2 + 25$.

a. $(4x - 5i)(x + 5i)$
b. $(4x + 5i)(x - 5i)$
c. $(2x - 5)(2x + 5)$
d. $(2x + 5i)(2x - 5i)$

25. Factor: $9x^2 + 42ix - 49$.

a. $(3x + 7i)^2$
b. $(3x - 7i)^2$
c. $(3x + 7i)(3x - 7i)$
d. $i(3x + 7)^2$

26. Which of the following is a true implication of the Fundamental Theorem of Algebra?

a. Every polynomial has both real and complex roots.
b. The discriminant determines the number of complex roots.
c. The degree and the number of roots of a polynomial are equal.
d. If a complex number is a root, its conjugate is also a root.

27. Find the magnitude and direction of $\overrightarrow{AB}$, assuming each gridline corresponds to 1 unit.

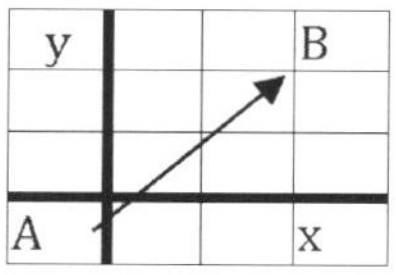

a. Direction: 2, Magnitude: 2
b. Direction: $\frac{2}{2}$, Magnitude: 1
c. Direction: 45°, Magnitude: $2\sqrt{2}$
d. Direction: 45°, Magnitude: 8

28. Which of the following vectors has a direction of 135°?

a. 

b. 

c. 

d. 

29. Find the magnitude and direction, respectively, of vector $\overrightarrow{\boldsymbol{MN}}$ with initial point M at (7,6) and endpoint N at (10,2).

a. 5, -53°
b. 5, -37°
c. 7, -53°
d. 7, -37°

30. Find the endpoint N of $\overrightarrow{MN}$. Let M = (1,1), $|\overrightarrow{MN}| = 5$, and the direction be 53°.

a. (3,4)
b. (4,3)
c. (4,5)
d. (5,4)

31. A plane is traveling at a constant air speed of 300 mph due north. Wind is blowing from a bearing of 50° east of north at 40 mph. Determine the actual speed and direction of the plane relative to the ground.

a. 327 mph, 5° east of north
b. 332 mph, 4° east of north
c. 340 mph, 5° east of north
d. 356 mph, 4° east of north

32. A boat is traveling back to the dock at 20 mph. The dock is 40° south of west from the boat's location. The current is going due west at 3 mph. What is the speed of the boat with respect to the water?

a. 20.2 mph
b. 19.6 mph
c. 18.4 mph
d. 17.8 mph

33. Which of the following is a correct depiction of $\vec{u} - \vec{v}$?

a. 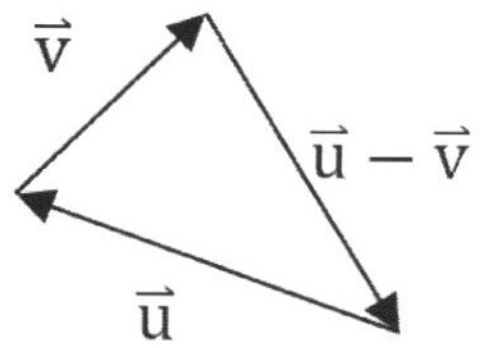

b. 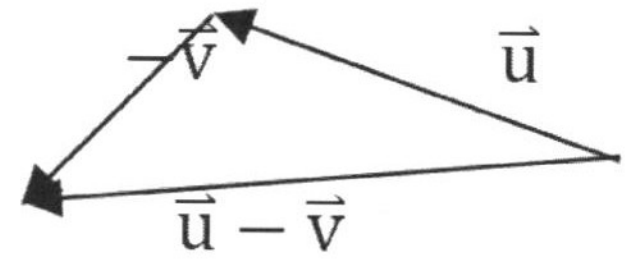

c. 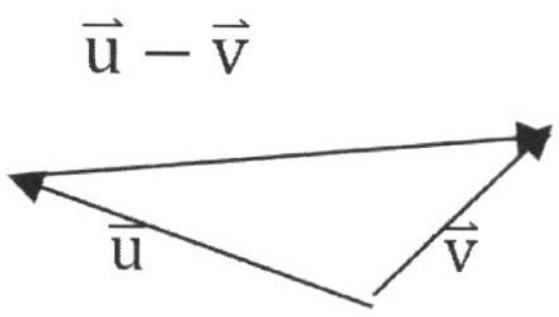

d. 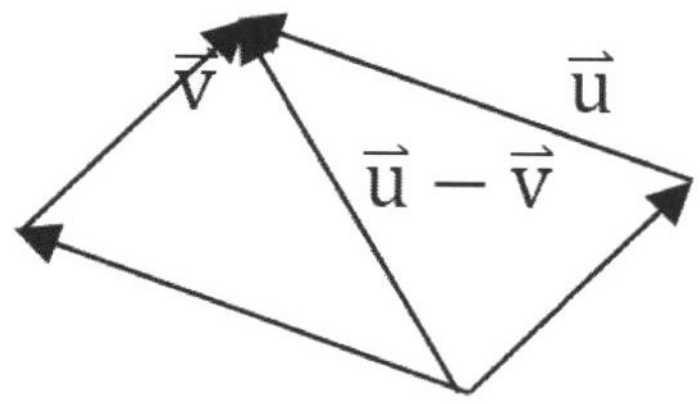

34. When is $|\vec{A} + \vec{B}| = |\vec{A}| + |\vec{B}|$?

a. When the directions of $\vec{A}$ and $\vec{B}$ are the same.
b. When the magnitude of $\vec{A}$ or $\vec{B}$ is equal to one.
c. When the directions of $\vec{A}$ and $\vec{B}$ are opposite.
d. When the angle between $\vec{A}$ and $\vec{B}$ is a quadrantal.

35. Define $\vec{u}$ to have magnitude of 4 and direction of 45° and $\vec{v}$ to have magnitude 10 and direction 150°. Find both the magnitude and direction of $\vec{u} + \vec{v}$.

a. 9.76, 82°
b. 9.76, 127°
c. 9.17, 105°
d. 9.17, 135°

36. Define $\vec{a}$ to have a magnitude of 4 and a direction of 75° and $\vec{b}$ to have a magnitude of 5 and a direction of 100°. Find both the magnitude and direction of $\vec{a} - \vec{b}$.

a. 4.75, -1°
b. 4.75, -29°
c. 2.18, -1°
d. 2.18, -29°

37. Which of the following is not a correct expression representing the pictured vector?

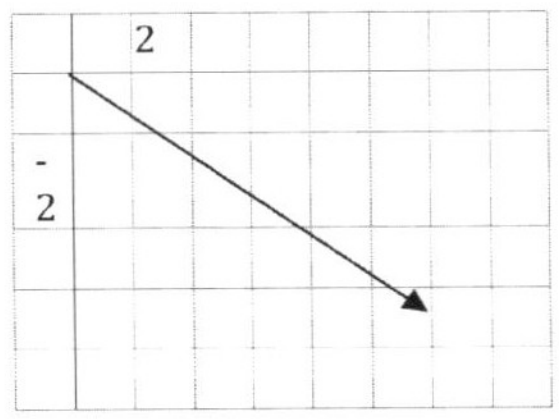

a. $-2(-6\hat{\imath} + 4\hat{\jmath})$
b. $4(3\hat{\imath} - 2\hat{\jmath})$
c. $-6\left(-2\hat{\imath} + \frac{4}{3}\hat{\jmath}\right)$
d. $8(\frac{3}{4}\hat{\imath} - \hat{\jmath})$

38. Compute $3\vec{a} - 2\vec{b}$ when $\vec{a} = -2\hat{\imath} + 7\hat{\jmath}$ and $\vec{b} = 4\hat{\imath} - 5\hat{\jmath}$.

a. $-14\hat{\imath} + 31\hat{\jmath}$
b. $-14\hat{\imath} + 11\hat{\jmath}$
c. $-11\hat{\imath} + 20\hat{\jmath}$
d. $2\hat{\imath} + 11\hat{\jmath}$

39. Compute the magnitude and direction of $3\vec{a}$. Let $\vec{a} = -7\hat{\imath} + 5\hat{\jmath}$. Round to the nearest integer.

a. 15, 36°
b. 26, 36°
c. 15, 144°
d. 26, 144°

40. The matrix represents the number of tickets sold on each day to each event. Which of the following statements can be made from the information in the matrix?

$$\begin{array}{r} \\ \textit{Football Game} \\ \textit{School Play} \end{array} \begin{array}{c} \begin{array}{ccc} \textit{Fri} & \textit{Sat} & \textit{Sun} \end{array} \\ \begin{bmatrix} 437 & 0 & 351 \\ 248 & 342 & 315 \end{bmatrix} \end{array}$$

a. The total number of football tickets sold was more than the total number of school play tickets.
b. The greatest number of tickets was sold at Friday's football game.
c. More tickets were sold on Sunday than on the other days.
d. There was a party after the play on Sunday.

41. The matrix below represents the price of gas at local stations. Change the matrix to reflect a 20% price increase at all stations for all octane ratings.

$$\begin{array}{r} \\ 77 \\ Chell \\ Arkco \end{array}\begin{array}{c} \begin{array}{ccc} regular & premium & super \end{array} \\ \begin{bmatrix} \$3.99 & \$4.09 & \$4.19 \\ \$4.02 & \$4.12 & \$4.22 \\ \$3.89 & \$3.99 & \$4.09 \end{bmatrix} \end{array}$$

a. $\begin{bmatrix} \$.80 & \$.82 & \$.83 \\ \$.80 & \$.82 & \$.84 \\ \$.78 & \$.80 & \$.81 \end{bmatrix}$

b. $\begin{bmatrix} \$4.01 & \$4.11 & \$4.21 \\ \$4.04 & \$4.14 & \$4.24 \\ \$3.91 & \$4.01 & \$4.11 \end{bmatrix}$

c. $\begin{bmatrix} \$4.19 & \$4.29 & \$4.39 \\ \$4.22 & \$4.32 & \$4.42 \\ \$4.09 & \$4.19 & \$4.29 \end{bmatrix}$

d. $\begin{bmatrix} \$4.79 & \$4.91 & \$5.03 \\ \$4.82 & \$4.94 & \$5.06 \\ \$4.67 & \$4.79 & \$4.91 \end{bmatrix}$

42. Solve: $\begin{bmatrix} 4 & 2 \\ 7 & 12 \end{bmatrix} + \begin{bmatrix} -1 & 15 \\ 3 & -5 \end{bmatrix}$.

a. $\begin{bmatrix} 3 & 17 \\ 10 & 7 \end{bmatrix}$

b. $\begin{bmatrix} 3 & 1 \\ 10 & 15 \end{bmatrix}$

c. $\begin{bmatrix} 19 & 17 \\ 2 & 7 \end{bmatrix}$

d. $\begin{bmatrix} 19 & 1 \\ 2 & 15 \end{bmatrix}$

43. Solve: $\begin{bmatrix} 75 \\ 55 \\ 37 \end{bmatrix} - 2\begin{bmatrix} 21 \\ 15 \\ 12 \end{bmatrix}$.

a. $\begin{bmatrix} -42 \\ -30 \\ -24 \end{bmatrix}$

b. $\begin{bmatrix} 54 \\ 40 \\ 25 \end{bmatrix}$

c. $\begin{bmatrix} 33 \\ 25 \\ 13 \end{bmatrix}$

d. $\begin{bmatrix} 33 \\ 40 \\ 25 \end{bmatrix}$

44. Evaluate: $\begin{bmatrix} 2 & 3 & 1 \\ 3 & 4 & 0 \end{bmatrix} \cdot \begin{bmatrix} 10 & 1 \\ 11 & 2 \\ 2 & 1 \end{bmatrix}$.

a. $\begin{bmatrix} 55 & 10 \\ 74 & 11 \end{bmatrix}$
b. $\begin{bmatrix} 20 & 3 \\ 33 & 8 \\ 2 & 0 \end{bmatrix}$
c. $\begin{bmatrix} 20 & 33 & 2 \\ 3 & 8 & 0 \end{bmatrix}$
d. $\begin{bmatrix} 55 & 10 \\ 76 & 12 \end{bmatrix}$

45. The dimensions of matrices A and B are 2x2, and matrix I is the 2x2 identity matrix. Which of the following is false?

a. A+B = B+A
b. k (A+B) = kA + kB
c. A·I = A
d. A·B = B·A

46. Find matrix B such that A·B=A, if A = $\begin{bmatrix} 15 & -37 \\ -28 & 41 \end{bmatrix}$.

a. $\begin{bmatrix} 0 & 0 \\ 0 & 0 \end{bmatrix}$
b. $\begin{bmatrix} 0 & 1 \\ 1 & 0 \end{bmatrix}$
c. $\begin{bmatrix} 1 & 0 \\ 0 & 1 \end{bmatrix}$
d. $\begin{bmatrix} 1 & 1 \\ 1 & 1 \end{bmatrix}$

47. Find the determinant of $A = \begin{bmatrix} 5 & 2 \\ 4 & 1 \end{bmatrix}$.

a. 18
b. 6
c. 3
d. -3

48. Multiply the matrix by the vector: $\begin{bmatrix} 2 & 1 & 3 \\ 1 & 1 & 3 \\ 3 & 2 & 1 \end{bmatrix} \cdot \begin{bmatrix} 10 \\ 8 \\ 5 \end{bmatrix}$

a. $\begin{bmatrix} 43 \\ 33 \\ 51 \end{bmatrix}$
b. $\begin{bmatrix} 43 \\ 23 \\ 46 \end{bmatrix}$
c. $\begin{bmatrix} 60 \\ 40 \\ 30 \end{bmatrix}$
d. $\begin{bmatrix} 16 \\ 13 \\ 11 \end{bmatrix}$

49. Determine the matrix M that would transform $\overrightarrow{a_1}$ into $\overrightarrow{a_2}$, by $M\overrightarrow{a_1} = \overrightarrow{a_2}$.

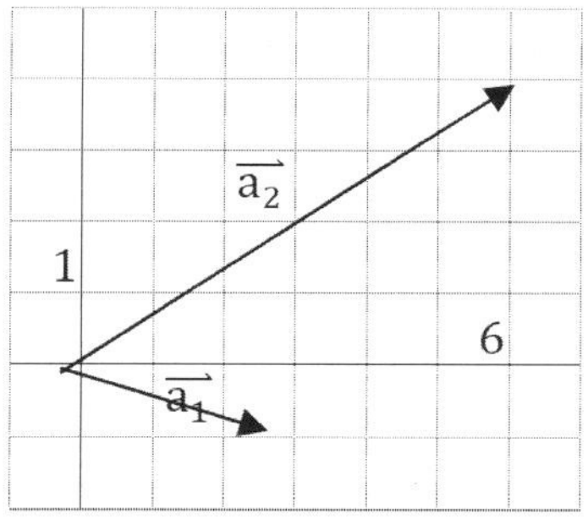

a. $\begin{bmatrix} 2 & 0 \\ 0 & -4 \end{bmatrix}$
b. $\begin{bmatrix} 0 & 2 \\ -4 & 0 \end{bmatrix}$
c. $\begin{bmatrix} 3 & 6 \\ -1 & 4 \end{bmatrix}$
d. $\begin{bmatrix} 6 & 3 \\ 4 & -1 \end{bmatrix}$

50. Describe the transformations in the plane that result from the following: $\begin{bmatrix} 2 & 0 \\ 0 & -2 \end{bmatrix} \cdot \vec{v}$.

a. $\vec{v}$ is translated 2 units to the left and to the right.
b. $\vec{v}$ is reflected across the origin and scaled by 2.
c. $\vec{v}$ is reflected across the x-axis and is doubled in length.
d. $\vec{v}$ is reflected across the y-axis and its magnitude doubles.

Answers and Explanations

Math.Content.N.RN.1

1. D: Use two properties of exponents: $b^x \cdot b^y = b^{x+y}$ and $(b^x)^y$. That is, the exponent on a base is the sum of the powers in a product (if the bases are the same), and the resulting power when an exponent is raised to a power is the product of the two exponents. Begin by adding the exponents within the parentheses. Next multiply that sum and the exponent outside the parentheses. $(5^{1/3} \cdot 5^2)^6 = (5^{7/3})^6 = 5^{14}$.

Math.Content.N.RN.1

2. C: Each of the statements follows an exponent rule. In A, the exponents inside and outside the parentheses are multiplied. $(8^{1/2})^2 = 8 = 2^3$. In B, the base is the same for each term in the product, so the exponents can all be added. $4^1 \cdot 4^1 \cdot 4^1 \cdot 4^1 \cdot 4^{1/2} = 4^{1+1+1+1+1/2} = 4^{9/2}$. In statement D, the base is the same for each term in the quotient, so the exponents can be subtracted. $\frac{\sqrt[6]{5}}{\sqrt[6]{5^2}} = \frac{5^{1/6}}{5^{2/6}} = 5^{-1/6}$. C follows the same rule as B, but there is a mistake: $5^{1/3} \cdot 5^{2/3} \cdot 5^{1/3} = 5^{4/3} = 5\sqrt[3]{5} \neq 5\sqrt{5}$.

Math.Content.N.RN.2

3. B: Fractional exponential notation can be obtained with the following identity: $\sqrt[y]{b^x} = b^{x/y}$. $5\sqrt[3]{2^2} + \sqrt[7]{3^2} = 5 \cdot 2^{2/3} + 3^{2/7}$. This cannot be simplified further by other rules governing exponentials because the bases are not the same.

Math.Content.N.RN.2

4. C: First write the radicals as fractional exponents on the same base. Next add the exponents in the numerator to write the product of those three terms. Finally subtract the exponent in the denominator and simplify:

$$\frac{5^8 \cdot \sqrt[3]{5^2} \cdot 5^{-3}}{\sqrt[3]{5^5}} = \frac{5^8 \cdot 5^{2/3} \cdot 5^{-3}}{5^{5/3}} = \frac{5^{17/3}}{5^{5/3}} = 5^{12/3} = 5^4$$

Math.Content.N.RN.3

5. A: Irrational numbers cannot be written as fractions. $\sqrt{30}$, $\sqrt[3]{150}$, and $(8)^{1/5}$ are all irrational numbers and remain irrational numbers after addition and multiplication (so that even though $8^{\frac{2}{3}}$ is rational, the sum is not). However, $\sqrt{121} = 11$ is rational and will remain rational after any operation with a rational number, including adding 7. (Further, a sum or product with of a rational number and an irrational number will be irrational.)

Math.Content.N.Q.1

6. A: Let the units guide the solution. First, create conversion fractions using the given information: 10 people = 11 sandwiches, and 20 pounds = 90 sandwiches. Then, start with the number of people and multiply by the fractions ensuring that the units will cancel as appropriate. The answer needs to be pounds of lunch meat so pounds should be the only remaining unit.

$$(300\ people) \cdot \frac{(11\ sandwiches)}{(10\ people)} \cdot \frac{(20\ pounds)}{(90\ sandwiches)}$$
$$= 300\ \cancel{people}) \cdot \frac{(11\ \cancel{sandwiches})}{(10\ \cancel{people})} \cdot \frac{(20\ pounds)}{(90\ \cancel{sandwiches})} = 73.333\ pounds$$

Derek needs to buy 73.333 pounds of lunch meat.

Math.Content.N.Q.2

7. C: The time period of one half-life is one week. It will take many half-lives for the mass to decay from 400 kilograms to 6.1 grams. At 10 weeks, for example, the mass is $400\left(\frac{1}{2}\right)^{10} = .3906\text{ kg} = 390.6\ grams$. Expressing time in units smaller than weeks will result in an unwieldy and unnecessarily finely-divided time axis. On the other hand, labeling the time axis in years does not provide enough divisions to accurately resolve individual half-lives. Labeling the axis in weeks is therefore the best choice.

Math.Content.N.Q.3

8. D: We could find Mr. Bolt's speed (the distance traveled divided by the elapsed time), but since the question only asks for who wins by how much time, and we already know how much time Mr. Bolt takes, we don't need to. To compare the two runners, then, we just have to convert the deer's speed from km/h to m/s, and calculate how many seconds the deer would take to run the 100 m. Set up conversion fractions so the units will cancel appropriately, and multiply them:

$$\frac{48\text{ km}}{\text{hr}} \cdot \frac{1000\text{ m}}{1\text{ km}} \cdot \frac{1\text{ hr}}{60\text{ min}} \cdot \frac{1\text{ min}}{60\text{ sec}} = \frac{48\ \cancel{\text{km}}}{\cancel{\text{hr}}} \cdot \frac{1000\text{ m}}{1\ \cancel{\text{km}}} \cdot \frac{1\ \cancel{\text{hr}}}{60\ \cancel{\text{min}}} \cdot \frac{1\ \cancel{\text{min}}}{60\text{ sec}} = 13.33\ m/s$$

The deer's speed is 13.33 m/s, and to find the deer's time, divide the length of the race by its speed: $\frac{100\text{ m}}{13.33\text{ m/s}} = 7.5$ seconds, and the deer wins! Compare the two times: 9.58 (Bolt) – 7.5 (Deer) = 2.08. The deer wins by 2.08 seconds.

Math.Content.N.Q.1

9. B: Use dimensional analysis to convert the given information to the number of pills needed. Construct conversion fractions based on the given equivalencies. For example, the statement "take 60 mg every 4 hours" corresponds to the fraction $\frac{60\text{ mg}}{4\text{ hours}}$. Use these fractions or their reciprocals to make sure the units cancel and leave the desired unit, pills:

$$5\text{ days} \cdot \frac{24\text{ hours}}{1\text{ day}} \cdot \frac{60\text{ mg}}{4\text{ hours}} \cdot \frac{1\text{ pill}}{30\text{ mg}} = 5\ \cancel{\text{days}} \cdot \frac{24\ \cancel{\text{hours}}}{1\ \cancel{\text{day}}} \cdot \frac{60\ \cancel{\text{mg}}}{4\ \cancel{\text{hours}}} \cdot \frac{1\text{ pill}}{30\ \cancel{\text{mg}}} = 60\text{ pills.}$$

Math.Content.N.Q.1

10. D: Since she buys one cookie for every student, the amount she spends depends on how many students there are; the cost is the dependent variable, dependent on the number of students, which is the independent variable. Choices C and D show this relationship with the independent variable on the x-axis and the dependent on the y-axis. The scale of the cost axis in choice C is too large, though: for 30 students Meghan only spends about $16. Since the scale of the cost axis in choice D is more appropriate, choice D is the best answer.

Math.Content.N.CN.1

11. C: To make applying the order of operations rules easier, rewrite the radical: $3 + (3^2 - 4 \cdot 5)^{\frac{1}{2}}$. First work what is inside the parentheses, treating the exponent first, then the multiplication, and then the subtraction: $3^2 - 4 \cdot 5 = 9 - 20 = -11$. Then combine, $3 + \sqrt{-11}$. Substitute $\sqrt{-1} = i$, and the most simplified form is $3 + i\sqrt{11}$.

Math.Content.N.CN.2

12. A: Multiply the terms of the complex binomials using the FOIL method. To FOIL, multiply the First terms, the Outer terms, the Inner terms, and the Last terms, then add all the products together. Next, replace $i^2 = -1$. Finally, combine like terms. $15 + 12i - 30i - 24i^{\ 2} = 15 + 12i - 30i + 24 = 39 - 18i$.

Math.Content.N.CN.2

13. B: Start by distributing the real numbers. Then combine like terms:

$$3(6 + 2i) - 4(1 + i) = 18 + 6i - 4 - 4i = 14 + 2i$$

Math.Content.N.CN.3

14. B: Multiply both numerator and denominator by the denominator's conjugate, which can be found by replacing *i* with *-i*. Use the FOIL to evaluate these products, remembering $i^2 = -1$, and combine like terms:

$$\frac{3 + 4i}{5 - 2i} \cdot \frac{5 + 2i}{5 + 2i} = \frac{15 + 6i + 20i + 8i^2}{25 - 4i^2} = \frac{15 + 26i - 8}{25 + 4} = \frac{7 + 26i}{29}$$

Math.Content.N.CN.3

15. C: The modulus can be found through the formula $|z|^2 = z \cdot \bar{z}$; that is, the square of the modulus of a complex number is equal to the product of that number and its complex conjugate, found by replacing *i* with *-i*:

$$|z|^2 = (6 + 8i)(6 - 8i) = 36 - 64i^2 = 36 + 64 = 100$$

Therefore, $|z| = \sqrt{100} = 10$.

Math.Content.N.CN.4

16. A: Every complex number, a+bi, is comprised of two parts: the real part is a, and the complex part is bi. The complex plane represents the real part, ℝ, on the horizontal axis and the coefficient of the complex part, ℂ, on the vertical. To graph a number on the complex plane, count "a" units to the right(left) for positive(negative) a on the real axis, then up(down) "b" units for positive(negative) b on the complex axis. Therefore, 3+2i is three units to the right and two units up from the origin. Choice A is the correct graph.

Math.Content.N.CN.4

17. D: The polar form of a complex number $z = a + bi$ is given by $z = r(cos\theta + isin\theta)$ where $r = \sqrt{a^2 + b^2}$ and $tan\theta = \frac{b}{a}$, so that $\theta = \tan^{-1}\frac{b}{a}$. In this case, $r = \sqrt{12^2 + 5^2} = 13$ and $\theta = tan^{-1}\frac{5}{12} = 22.6 \approx 23°$. Substituting r and θ into polar form, $z = 13(cos23° + isin23°)$.

Math.Content.N.CN.5

18. A: Vectors can be graphically added using the parallelogram method, in which the diagonal of the parallelogram, formed by duplicating each of the two vectors originating at the tip of the other, represents the sum. Complex numbers are of the form $a + bi$; a is an element of the real numbers and is represented on the horizontal axis, bi is an element of the complex numbers and b is represented on the vertical axis. Find the complex numbers representing the legs of the parallelogram by simply counting the horizontal and vertical components; they are $-3 + 2i$ and $4 + 2i$. The diagonal of the parallelogram, the result, is again found by counting: $(-3 + 2i) + (4 + 2i) = 1 + 4i$.

Math.Content.N.CN.5

19. D: It is generally easiest to multiply complex numbers in polar form. So, first convert each number to polar form: $a + bi = r(\cos\theta + i\sin\theta)$ where $r = \sqrt{a^2 + b^2}$ and $\theta = \tan^{-1}\frac{b}{a}$. Thus, for $12 + 5i, r = \sqrt{12^2 + 5^2} = 13, \theta = \tan^{-1}\frac{5}{12} \approx 23°$ so $12 + 5i = 13(\cos 23° + i\sin 23°)$. For $-4 + 3i$, $r = \sqrt{(-4)^2 + 3^2} = 5, \theta = \tan^{-1}\frac{3}{-4} \approx -37°$. However, since this point is in the second quadrant, the direction is found by $180 + \theta = 143°$, and $-4 + 3i = 5(\cos 143° + i\sin 143°)$. To multiply complex numbers in polar form, multiply the radii and add the angular directions. Therefore,

$$(13(\cos 23 + i\sin 23))(5(\cos 143 + i\sin 143)) = (5 \cdot 13)[\cos(23 + 143) + i\sin(23 + 143)]$$

$$= 65(\cos 166 + i\sin 166)$$

Math.Content.N.CN.6

20. B: To find the midpoint between two complex numbers, add the complex numbers and divide by 2: $\frac{(-30+15i)+(12-3i)}{2} = \frac{-18+12i}{2} = -9 + 6i$.

Math.Content.N.CN.6

21. C: Distances on the complex plane are calculated very much like they are on the real plane. To find the distance, use the distance formula $D = \sqrt{(x_1 - x_2)^2 + (y_1 - y_2)^2}$ with $a = x$ and $b = y$:

$$D = \sqrt{((-30) - 12)^2 + (15 - (-3))^2} = \sqrt{(-42)^2 + 18^2} \approx 46$$

Math.Content.N.CN.7

22. C: Set the equation equal to zero. The calculation is easier if the coefficient of the x-squared term is positive: $2x^2 - 5x + 11 = 0$. Now that the equation is in the form $ax^2 + bx + c$, we can substitute the coefficients into the quadratic formula:

$$x = \frac{-b\pm\sqrt{b^2-4ac}}{2a} = \frac{5\pm\sqrt{5^2-4\cdot(2)\cdot(11)}}{2\cdot(2)} = \frac{5\pm\sqrt{25-88}}{4} = \frac{5\pm\sqrt{-63}}{4} = \frac{5\pm 3i\sqrt{7}}{4}.$$

Math.Content.N.CN.7

23. A: The equation is already in standard form, $y = ax^2 + bx + c$. By substituting the coefficients into the quadratic formula, we get $x = \frac{-b\pm\sqrt{b^2-4ac}}{2a} = \frac{2\pm\sqrt{(-2)^2-4\cdot1\cdot10}}{2\cdot1} = \frac{2\pm\sqrt{4-40}}{2} = \frac{2\pm\sqrt{-36}}{2} = \frac{2\pm 6i}{2} = 1 \pm 3i$.

Math.Content.N.CN.8

24. D: A difference of two perfect squares can be factored as: $a^2 - b^2 = (a+b)(a-b)$. When asked to factor the sum of squares, apply the difference of squares rule as follows: $a^2 + b^2 = a^2 - (-b^2) = (a \pm \sqrt{-b^2})(a \mp \sqrt{-b^2})$. Alternatively, $a^2 + b^2 = a^2 - b^2 i^2 = a^2 - (bi)^2 = (a+bi)(a-bi)$. In this case, then: $4x^2 + 25 = 4x^2 - 25i^2 = (2\text{x})^2 - (5i)^2 = (2x - 5i)(2x + 5i)$.

Math.Content.N.CN.8

25. A: A perfect square trinomial is one that can be written as the square of a binomial as follows: $a^2x^2 + 2abx + b^2 = (ax+b)^2$. Notice that $42i = 2 \cdot (3 \cdot 7 \cdot i\ \ )$ and apply the pattern: $9x^2 + 42ix - 49 = (3x)^2 + 42ix + (7i)^2 = (3x + 7i)^2$.

Math.Content.N.CN.9

26. C: Choice A is clearly false, since some polynomials only have real roots. While the discriminant determines whether there are complex roots, if there is one complex root its conjugate is also a root (choice D is true, then, but not a consequence of the Fundamental Theorem). The Fundamental Theorem of Algebra states that a polynomial *P(z)* of degree *n* has *n* values z_i for which $P(z_i) = 0$. In other words, the degree and the number of roots are equal in every polynomial, and C is correct.

Math.Content.N.VM.1

27. C: The direction is found by measuring the angle between the x-axis and the vector or with the tangent function, $\theta = \tan^{-1}\frac{y}{x}$, where *y* is the vertical component of the vector, *x* is the horizontal component of the vector: $\theta = \tan^{-1}\frac{2}{2} = 45°$. The magnitude, $|\overrightarrow{AB}|$, is the distance from the initial point to the endpoint. In general it can be found by applying the distance formula, but in this case it can be found using the simpler Pythagorean Theorem, $|\overrightarrow{AB}|^2 = x^2 + y^2$. In this case, $|\overrightarrow{AB}| = \sqrt{2^2 + 2^2} = \sqrt{8} = 2\sqrt{2}$.

Math.Content.N.VM.1

28. D: The direction of a vector is found by measuring the angle between the positive x-axis and the vector. Since 90 < 135 < 180, this angle corresponds to a vector in the second quadrant that starts at the origin. Choice D is correct.

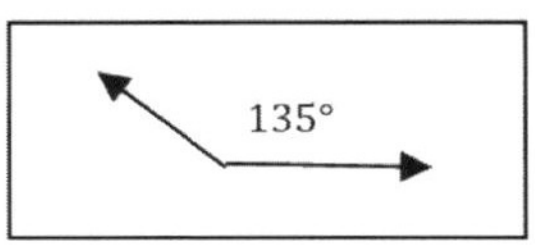

Math.Content.N.VM.2

29. A: The magnitude can be found using the distance formula: $|\overrightarrow{MN}| = \sqrt{(x_1 - x_2)^2 + (y_1 - y_2)^2} = \sqrt{(7-10)^2 + (6-2)^2} = \sqrt{9+16} = 5$. Plot the points to know which quadrant the answer is in. If the vector points anywhere other than Quadrant I or IV, θ will be a reference angle, and not the exact answer. As this vector points to Q4, θ will be negative.

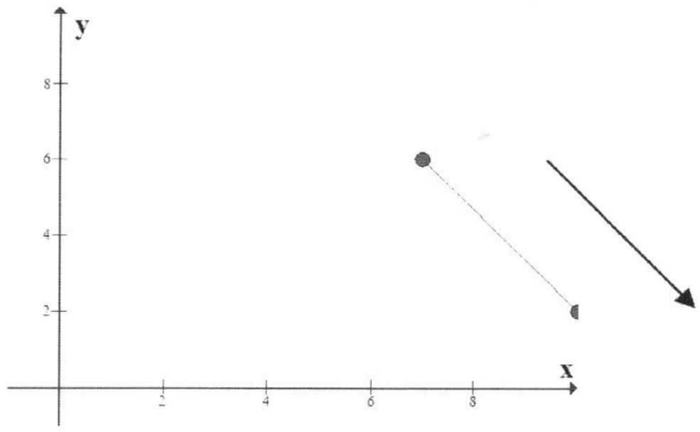

The direction, θ, is found using the inverse tangent function:

$$\theta = \tan^{-1}\left(\frac{y_2-y_1}{x_2-x_1}\right) = \tan^{-1}\left(\frac{2-6}{10-7}\right) = \tan^{-1}\frac{4}{-3} = -53°.$$

Math.Content.N.VM

30. C: Using the information given, draw a sketch of the problem.

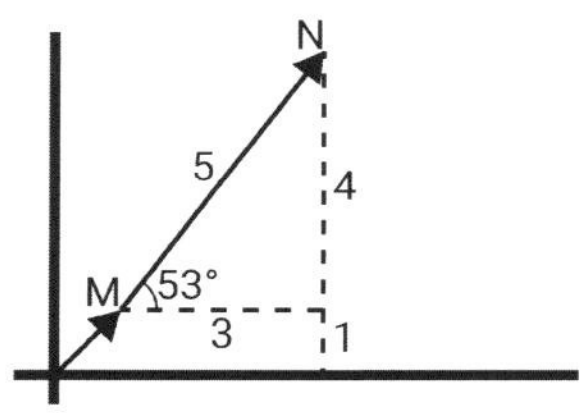

Use the trigonometric functions to find the sides of the triangle:

$$\cos 53° = \frac{x}{5} \xrightarrow{yields} x = 5 \cdot \cos 53 \approx 3 \quad \text{and} \quad \sin 53° = \frac{y}{5} \xrightarrow{yields} 5 \cdot \sin 53 \approx 4$$

The initial point of $\overrightarrow{MN}$ is (1,1), so the endpoint is (1+3, 1+4) = (4,5).

Math.Content.N.VM.3

31. A: Using the information given, sketch the problem, labeling the velocity of the plane $\vec{p}$, the velocity of the wind $\vec{w}$, and the velocity of the plane relative to the ground$\vec{v} = \vec{p} + \vec{w}$. The plane is flying due north, so its horizontal component is 0 and its vertical component equals its magnitude: $\vec{p} = 0\hat{\imath} + 300\hat{\jmath}$. Next find the horizontal and vertical components of the wind's velocity. The angle inside the triangle is 90° – 50° = 40°, so that $\sin 40 = \frac{w_y}{40} \xrightarrow{yields} w_y = 40 \sin 40 \approx 26$, and $\cos 40 = \frac{w_x}{40} \xrightarrow{yields} w_x = 40 \cos 40 \approx 31$. Therefore, $\vec{v} = (0\hat{\imath} + 300\hat{\jmath}) + (31\hat{\imath} + 26\hat{\jmath}) = 31\hat{\imath} + 326\hat{\jmath}$. The speed of the plane is the magnitude of this vector, $|\vec{v}| = \sqrt{31^2 + 326^2} \approx 327$, and the direction is given by $\theta = \tan^{-1}\frac{326}{31} \approx 85°$, although this is measured as north of east and the answer shoiuld be reported as 5° east of north.

Math.Content.N.VM.3

32. D: The speed of the boat with respect to the water can be found by subtracting the velocity of the current, $\vec{c}$, from the velocity of the boat, $\vec{b}$. First, find the components for each vector with basic trigonometric functions. The boat's direction is 40° south of west, which puts the vector into quadrant III as shown here

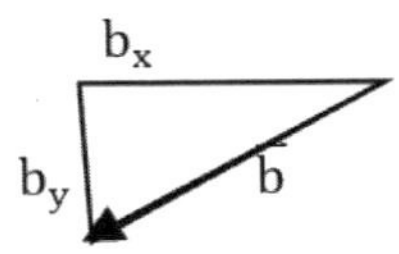

and makes our angle 40+180= 220°. Then we have: $\cos 220 = \frac{b_x}{20} \xrightarrow{yields} b_x = 20\cos 220$, and $\sin 220 = \frac{b_y}{20} \xrightarrow{yields} b_y = 20\sin 220 \approx -12.86$. So, $\vec{b} = -15.32\hat{\imath} - 12.86\hat{\jmath}$. The direction of the current is due west, so the vertical component of that velocity is zero and $\vec{c} = -3\hat{\imath}$. The speed relative to the water is equal to the magnitude of the difference:

$$|\vec{b} - \vec{c}| = |(-15.32\hat{\imath} - 12.28\hat{\jmath} + 3\hat{\imath})| = |(-12.32\hat{\imath} - 12.28\hat{\jmath})| = \sqrt{(-12.32)^2 + (-12.86)^2} = 17.8.$$

Math.Content.N.VM.4c

33. B: Vector addition and subtraction can be drawn end to end or following the parallelogram rule. To use the parallelogram rule for subtraction, add the negative of the subtracted vector; that is, add a vector of the same magnitude pointing in the opposite direction. For $\vec{u} - \vec{v}$, first draw $\vec{u}$. Next draw $-\vec{v}$, which is the same length as and antiparallel to $\vec{v}$, with its initial point at the endpoint of $\vec{u}$. Finally, draw the difference vector from the initial point of $\vec{u}$ to the endpoint of $-\vec{v}$, as in the parallelogram method; choice B correctly represents this.

Math.Content.N.VM.4a

34. A: The question asks when the magnitude of the sum of two vectors is the same as the sum of the magnitude of the two vectors. In other words, under what conditions will the directions of the two vectors not affect the sum of the magnitudes? This will only occur when the direction of both vectors is the same.

Math.Content.N.VM.4b

35. B: With the given information, sketch the vector sum and find as many angles in the resulting triangle as possible. Next, redraw the triangle with only the essential information.

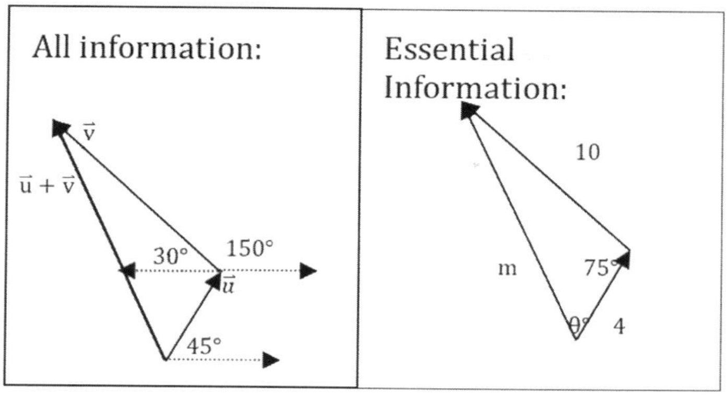

To find m, use the Law of Cosines: $m^2 = 4^2 + 10^2 - 2 \cdot 4 \cdot 10 \cdot cos75$, and $m = 9.76$. To find θ, use the Law of Sines: $\frac{sin75}{9.76} = \frac{sin\theta}{10}$, and $\sin\theta = \frac{10 \cdot sin75}{9.76}, \theta = \sin^{-1}.9896 \approx 82°$. Finally, find the direction, 82+45=127°.

Alternatively, express the vectors in terms of unit vectors: $\vec{u} = (4\cos 45)\hat{\imath} + (4\sin 45)\hat{\jmath} = 2.83\hat{\imath} + 2.83\hat{\jmath}$, and $\vec{v} = (10\cos 150)\hat{\imath} + (10\sin 150)\hat{\jmath} = -8.66\hat{\imath} + 5\hat{\jmath}$. Then add the vectors, $\vec{u} + \vec{v} = (2.83 - 8.66)\hat{\imath} + (2.83 + 5)\hat{\jmath} = -5.83\hat{\imath} + 7.83\hat{\jmath}$. Then, find the magnitude and direction: $|\vec{u} + \vec{v}| = \sqrt{(-5.83)^2 + 7.83^2} = 9.76$, and $\theta = \tan^{-1}\frac{7.83}{-5.83} = -53.3°$, but since the vector sum points towards the second quadrant, the direction is -53.3° + 180° = 126.7° ≈ 127°.

Math.Content.N.VM.4c

36. D: With the given information sketch the vector difference and determine as many angles as possible. Then draw the triangle with only the essential information.

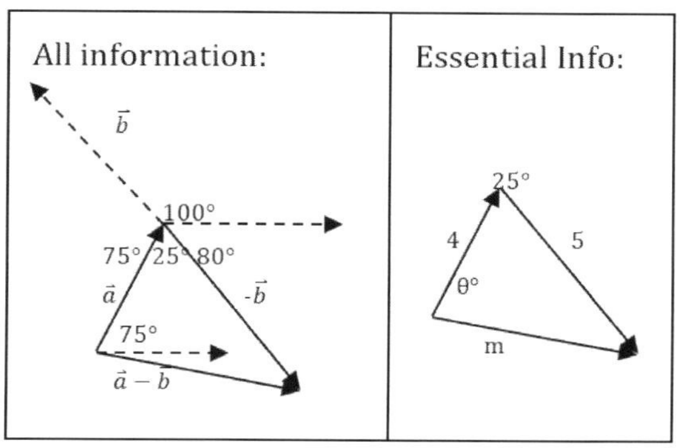

To find m, use the law of Cosines: $m^2 = 4^2 + 5^2 - 2 \cdot 4 \cdot 5\cos 25 \xrightarrow{yields} m = 2.18$. To find θ, use the Law of Sines: $\frac{sin25}{2.18} = \frac{sin\theta}{5} \xrightarrow{yields} sin\theta = \frac{5 \cdot sin25}{2.18} \xrightarrow{yields} \theta = \sin^{-1} 0.9693 \approx 76°\, or\ 104°$, and since you are looking for an obtuse angle, the correct value is 104°. Thus, the direction of $\vec{a} - \vec{b}$ is the direction of $\vec{a}$ minus θ, 75° - 104° = -29°.

Alternatively, express the vectors in terms of unit vectors: $\vec{a} = (4\cos 75)\hat{\imath} + (4\sin 75)\hat{\jmath} = 1.04\hat{\imath} + 3.86\hat{\jmath}$, and $\vec{b} = (5\cos 100)\hat{\imath} + (5\sin 100)\hat{\jmath} = -0.87\hat{\imath} + 4.92\hat{\jmath}$. Subtract the vectors component-wise: $\vec{a} - \vec{b} = 1.91\hat{\imath} - 1.06\hat{\jmath}$. Find the magnitude and direction: $|\vec{a} - \vec{b}| = \sqrt{1.91^2 + (-1.06)^2} = 2.18$, and $\theta = \tan^{-1}\frac{-1.06}{1.91} = -29°$.

Math.Content.N.VM.5a

37. D: A common notation for vectors separates the vector into the horizontal component, $\hat{\imath}$, and the vertical component, $\hat{\jmath}$. The graphed vector is $12\hat{\imath} - 8\hat{\jmath}$. After applying the distributive property, choices A, B, and C are all $12\hat{\imath} - 8\hat{\jmath}$. D is something different and therefore, not a correct expression for the graphed vector.

Math.Content.N.VM.5a

38. A: First substitute the value of $\vec{a}$ and $\vec{b}$ into $3\vec{a} - 2\vec{b}$. Distribute, then combine like terms: $\mathbf{3(-2\hat{\imath} + 7\hat{\jmath}) - 2(4\hat{\imath} - 5\hat{\jmath}) = -6\hat{\imath} + 21\hat{\jmath} - 8\hat{\imath} + 10\hat{\jmath} = -14\hat{\imath} + 31\hat{\jmath}}$.

Math.Content.N.VM.5b

39. D: Find $3\vec{a}$ by distributing: $3\vec{a} = 3(-7\hat{\imath} + 5\hat{\jmath}) = -21\hat{\imath} + 15\hat{\jmath}$. The magnitude is found with Pythagorean Theorem or the Distance Formula: $|3\vec{a}| = \sqrt{(-21)^2 + 15^2} = \sqrt{666} \approx 26$.

The direction of $3\vec{a}$ is the same as the direction of $\vec{a}$. To find the direction, determine the quadrant of the vector, then use the tangent function to find the reference angle.

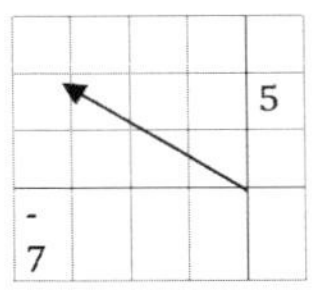

$\theta = tan^{-1}\frac{5}{-7} \approx -36$, but θ is a reference angle in the second quadrant, so the direction is given by 180 + θ = 144°.

Math.Content.N.VM.6

40. B: That the greatest number of tickets was sold at Friday's football game is the only true statement based on the matrix. A is false because the total football tickets was 788, as compared to 905 total play tickets. C is false as Friday has 19 more tickets sold than on Sunday. We have no information about parties so the truth of D cannot be found from this matrix.

Math.Content.N.VM.7

41. D: This is solved by scalar multiplication of the matrix. To increase each number by 20%, multiply each number in the matrix by 1.2 (100%+20% and converted to a decimal):

$$\mathbf{1.2\begin{bmatrix} 3.99 & 4.09 & 4.19 \\ 4.02 & 4.12 & 4.22 \\ 3.89 & 3.99 & 4.09 \end{bmatrix} = \begin{bmatrix} 1.2\cdot 3.99 & 1.2\cdot 4.09 & 1.2\cdot 4.19 \\ 1.2\cdot 4.02 & 1.2\cdot 4.12 & 1.2\cdot 4.22 \\ 1.2\cdot 3.89 & 1.2\cdot 3.99 & 1.2\cdot 4.09 \end{bmatrix} = \begin{bmatrix} 4.79 & 4.91 & 5.03 \\ 4.82 & 4.94 & 5.06 \\ 4.67 & 4.79 & 4.91 \end{bmatrix}}$$

Math.Content.N.VM.8

42. A: Matrices can be added or subtracted only if they have the same dimensions. Since these two matrices do, each position of the resulting matrix is the sum of the values of each matrix at the corresponding positions:

$$\begin{bmatrix} 4 & 2 \\ 7 & 12 \end{bmatrix} + \begin{bmatrix} -1 & 15 \\ 3 & -5 \end{bmatrix} = \begin{bmatrix} 4-1 & 2+15 \\ 7+3 & 12-5 \end{bmatrix} = \begin{bmatrix} 3 & 17 \\ 10 & 7 \end{bmatrix}$$

Math.Content.N.VM.7

43. C: First, make sure the dimensions are the same. Second, apply the scalar multiplication. Finally, subtract the elements in corresponding positions:

$$\begin{bmatrix} 75 \\ 55 \\ 37 \end{bmatrix} - 2\begin{bmatrix} 21 \\ 15 \\ 12 \end{bmatrix} = \begin{bmatrix} 75 \\ 55 \\ 37 \end{bmatrix} - \begin{bmatrix} 42 \\ 30 \\ 24 \end{bmatrix} = \begin{bmatrix} 33 \\ 25 \\ 13 \end{bmatrix}$$

Math.Content.N.VM.8

44. A: First check that matrix multiplication is possible: matrix multiplication is only possible when the number of columns in matrix A is equal to the number of rows in matrix B. Since the two matrices are of dimension (2×3) and (3×2), matrix multiplication is possible (the inner pair of dimensions are equal) and the dimension of the result is 2×2 (the outer pair). To find an element of the product, multiply each entry of the corresponding row of matrix A by the respective entry (in order) of the corresponding column of matrix B and add the products. The visual explanation is easier than the verbal; the first row $(2 \;\; 3 \;\; 1)$ is multiplied by and added to the first column $\begin{pmatrix} 10 \\ 11 \\ 2 \end{pmatrix}$, from which the element in the first row of the first column in final matrix results $(2 \times 10 + 3 \times 11 + 1 \times 2 = 55)$, then the first row is "multiplied by and added to" the second column, from which the element in the first row of the second column in final matrix results, and so on:

$$\begin{bmatrix} 2 & 3 & 1 \\ 3 & 4 & 0 \end{bmatrix} \times \begin{bmatrix} 10 & 1 \\ 11 & 2 \\ 2 & 1 \end{bmatrix} = \begin{bmatrix} 2 \times 10 + 3 \times 11 + 1 \times 2 & 2 \times 1 + 3 \times 2 + 1 \times 2 \\ 3 \times 10 + 4 \times 11 + 0 \times 2 & 3 \times 1 + 4 \times 2 + 0 \times 1 \end{bmatrix} \rightarrow \begin{bmatrix} 55 & 10 \\ 74 & 11 \end{bmatrix}$$

Math.Content.N.VM.9

45. D: A is true as matrix addition is commutative. B is true as the distributive property holds for scalar multiplication. C is the definition of the identity matrix. Matrix multiplication is not commutative, however, so D is false.

Math.Content.N.VM.10

46. C: This question asks for the identity matrix since it is defined as a matrix for which the product of any matrix with it is that matrix: $I = \begin{bmatrix} 1 & 0 \\ 0 & 1 \end{bmatrix}$. To verify this:

$$\begin{bmatrix} 15 & -37 \\ -28 & 41 \end{bmatrix} \cdot \begin{bmatrix} 1 & 0 \\ 0 & 1 \end{bmatrix} = \begin{bmatrix} 15 \cdot 1 + (-37) \cdot 0 & 15 \cdot 0 + \cdot (-37) \cdot 1 \\ (-28) \cdot 1 + (41) \cdot 0 & (-28) \cdot 0 + 41 \cdot 1 \end{bmatrix} = \begin{bmatrix} 15 & -37 \\ -28 & 41 \end{bmatrix}$$

Math.Content.N.VM.10

47. D: The determinant of $A = \begin{bmatrix} a & b \\ c & d \end{bmatrix}$ is $|A| = ad - bc$. In this case, $|\boldsymbol{A}| = \begin{vmatrix} 5 & 2 \\ 4 & 1 \end{vmatrix} = 5 \cdot 1 - 2 \cdot 4 = -3$.

Math.Content.N.VM.11

Math.Content.N.VM.11

48. A: For two matrices of dimension $m \times n$ and $n \times p$, since the number of columns of the first equals the number of rows of the second, the product has dimension $m \times p$, 3×1 in this case. To find an element of the product, multiply each entry of the corresponding row of matrix A by the respective entry (in order) of the corresponding column of matrix B and add the products:

$$\begin{bmatrix} 2 & 1 & 3 \\ 1 & 1 & 3 \\ 3 & 2 & 1 \end{bmatrix} \times \begin{bmatrix} 10 \\ 8 \\ 5 \end{bmatrix} = \begin{bmatrix} 2 \times 10 + 1 \times 8 + 3 \times 5 \\ 1 \times 10 + 1 \times 8 + 3 \times 5 \\ 3 \times 10 + 2 \times 8 + \times 1 \times 5 \end{bmatrix} = \begin{bmatrix} 43 \\ 33 \\ 51 \end{bmatrix}$$

Math.Content.N.VM.11

49. A: We need to find the matrix that transforms $\begin{bmatrix} 3 \\ -1 \end{bmatrix}$ into $\begin{bmatrix} 6 \\ 4 \end{bmatrix}$. The horizontal component is stretched by a factor of 2. The vertical component is reflected across the x-axis and stretched by a factor of 4. Using this information, write the matrix $\begin{bmatrix} 2 & 0 \\ 0 & -4 \end{bmatrix}$. Test the matrix to make sure it is correct: $\begin{bmatrix} 2 & 0 \\ 0 & -4 \end{bmatrix} \cdot \begin{bmatrix} 3 \\ -1 \end{bmatrix} = \begin{bmatrix} 2 \cdot 3 + 0 \cdot (-1) \\ 0 \cdot 3 + (-4) \cdot (-1) \end{bmatrix} = \begin{bmatrix} 6 \\ 4 \end{bmatrix}$.

Math.Content.N.VM.12

50. C: The matrix $\begin{bmatrix} 2 & 0 \\ 0 & -2 \end{bmatrix}$ will transform $\vec{v}$ in two ways. The numbers determine one transformation and the negative sign determines the second transformation. The 2's along the diagonal will double the length of the vector, while the negative in this position will reflect $\vec{v}$ across the y-axis.

Practice Test #2

Practice Questions

1. Simplify the expression: $\sqrt[5]{(4^3 \cdot 4^3)^2}$.

a. $4^{8/5}$
b. $4^{12/5}$
c. $4^{18/5}$
d. 4^{13}

2. Which of the following statements is false?

a. $^{a.}\sqrt[5]{21}$ is both real and irrational.
b. The sum of $\sqrt[5]{21}$ and a non-zero real number will always be an irrational number.
c. The difference of $\sqrt[5]{21}$ and a non-zero real number not equal to $\sqrt[5]{21}$ will always be an irrational number.
d. The product of $\sqrt[5]{21}$ and a non-zero real number will always be an irrational number.

3. Write the below expression in simplest form using fractional exponents.

$$\frac{3 \cdot 3 \cdot \sqrt[3]{3 \cdot 3} \cdot 5 \cdot 5 \cdot \sqrt[3]{7 \cdot 7 \cdot 7 \cdot 7}}{3 \cdot \sqrt[3]{5 \cdot 5} \cdot 7\sqrt[3]{7 \cdot 7}}$$

a. $3^{2/3} \cdot 5^{-1/3} \cdot 7^{-1/3}$
b. $3^{4/3} \cdot 5^2 \cdot 7^{2/3}$
c. $3^{5/3} \cdot 5^{4/3} \cdot 7^{-1/3}$
d. $3^{8/3} \cdot 5^2 \cdot 7^{4/3}$

4. Simplify: $\left(7^{12} \cdot 7^{-5}\right)^{2/3}$.

a. $7^{14/3}$
b. $7^{9/3}$
c. $7^{14/21}$
d. $7^{34/3}$

5. Which of the following is a rational number?

a. $5\sqrt[3]{49} + 1$
b. $\sqrt{99} \cdot \sqrt{7} + 3$
c. $3\pi - 4$
d. $2\sqrt{169}$

6. Noelle is a 45-pound 5-year-old girl who has been prescribed amoxicillin for an acute ear infection. The amoxicillin dosage for children aged 4 to 6 years is 90 mg per kg of body weight per day for 10 days. Knowing that 1kg = 2.2 lb., how much of the antibiotic will be needed to fill the prescription?

a. 18.4 grams
b. 19.8 grams
c. 40.5 grams
d. 89.1 grams

7. Three friends are driving from WSU, in Pullman, WA, to Gonzaga, in Spokane, WA. The distance is 77 miles. The car gets 23 mpg and has a 15 gallon tank. Gas costs $3.89/gallon. What will each friend's share of the gas be round trip, assuming they split the costs evenly?

a. $4.34
b. $8.68
c. $13.02
d. $19.45

8. Heidi owns a bakery and is converting her mother's soda bread recipe for commercial use. The recipe calls for 1 tablespoon of baking soda and makes 2 loaves of bread. She needs the recipe to make 350 loaves of bread. What is an appropriate unit of measure for baking soda in the converted recipe?

a. teaspoons
b. tablespoons
c. quarts
d. gallons

9. The average number of students per computer in US public schools can be estimated by the sequence model $a_n = 35.8127(0.8646)^n$, for n=1,2,3,...25, with the 1987-1988 school year corresponding to n=1. Choose the most appropriate scale for a graph of this model from 1987 to 2012.

a.

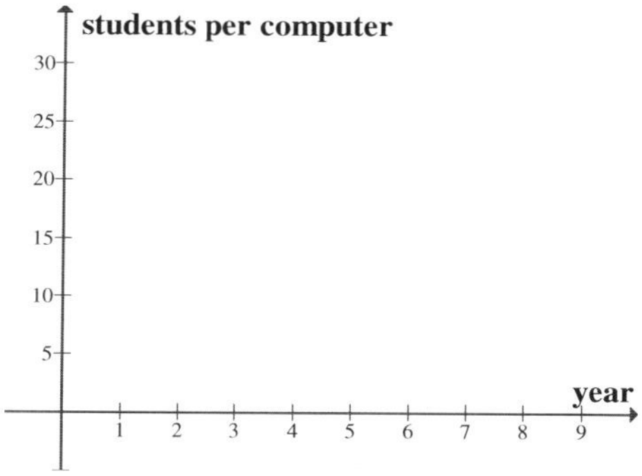

b.

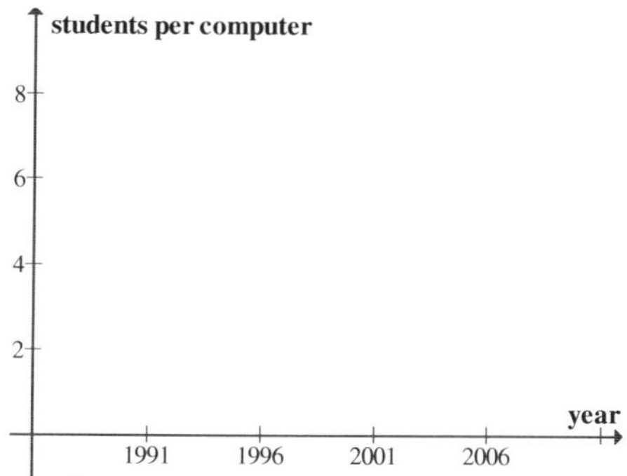

c.

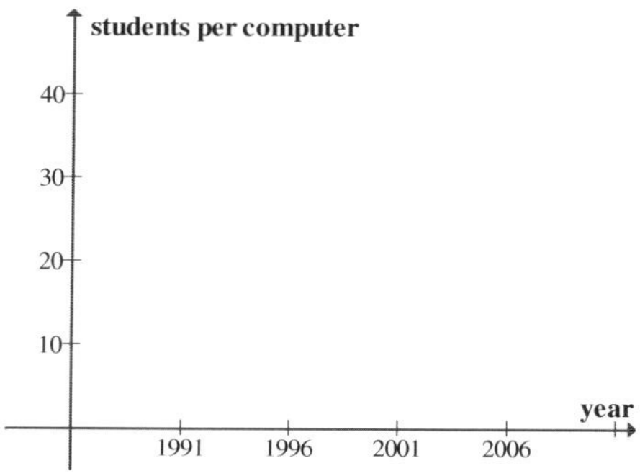

d.

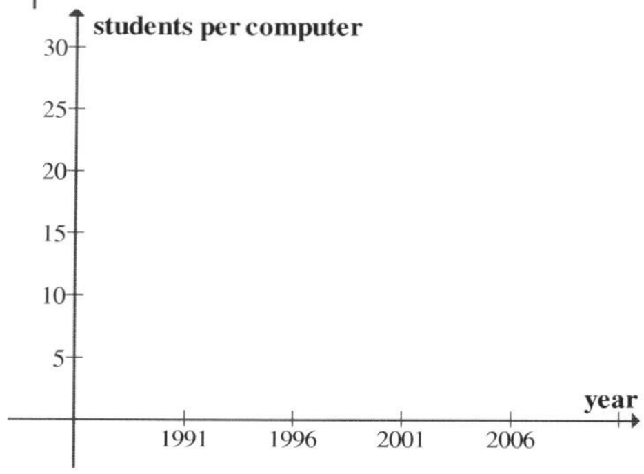

10. The expression $[(3.35 \cdot 10^{18})t^2]^{1/3}$ gives the average radius, *r*, of a planet's path around the sun in meters, where *t* is the time in Earth-seconds that it takes the planet to go around the sun. Choose an appropriate unit for reporting the average radius of the earth's path around the sun.

a. meters
b. kilometers
c. megameters
d. gigameters

11. Simplify: $3\sqrt{4 - 5 \cdot 8} + \sqrt{5 - 6}$.

a. 19i
b. 10i
c. 6i + 1
d. 10

12. Simplify: $4(-3 + 5i) + 6(1 - 3i)$.

a. 18 + 38i
b. 13 + 18i
c. −6 + 2i
d. −6 − 2i

13. Simplify: $(9 - 3i)(-7 - 12i)$.

a. $-99 - 87i$
b. $-27 - 129i$
c. $45 - 15i$
d. $2 - 15i$

14. Find the conjugate, $\bar{z}$, of $z = (8 - 6i)$.

a. $-8 + 6i$
b. $8 + 6i$
c. $28 - 96i$
d. 28

15. Evaluate the quotient and express in standard form: $\frac{15-5i}{4-2i}$.

a. $\frac{25+5i}{6}$
b. $\frac{7+5i}{2}$
c. $\frac{5+i}{2}$
d. $\frac{7+i}{2}$

16. Determine the polar form of the number graphed on the complex plane.

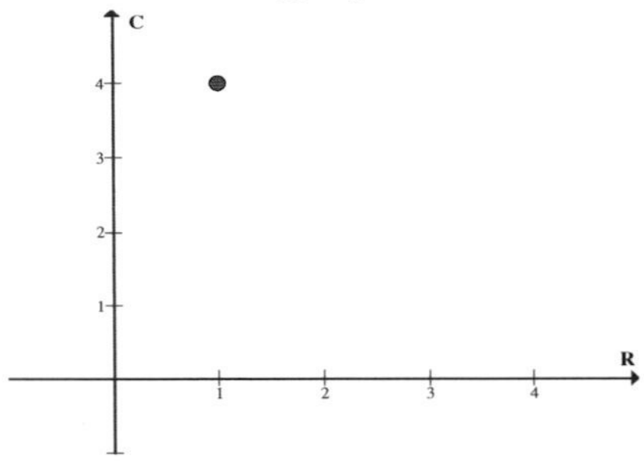

a. $\sqrt{10}(\cos 72 + i \sin 72)$
b. $\sqrt{17}(\cos 76 + i \sin 76)$
c. $4(\cos 80 + i \sin 80)$
d. $\sqrt{17}(\cos 14 + i \sin 14)$

17. Compute the distance between $(3 - 6i)$ and $(-2 + 10i)$ on the complex plane.

a. $\sqrt{17}$
b. $\sqrt{41}$
c. 16.76
d. 21

18. Which graph shows a conjugate pair?

a. 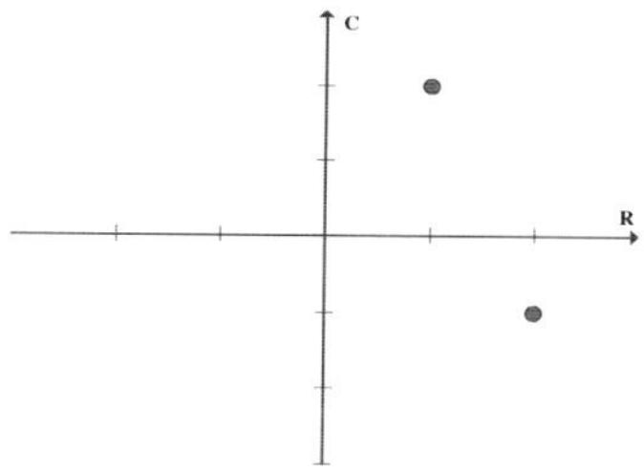

b. 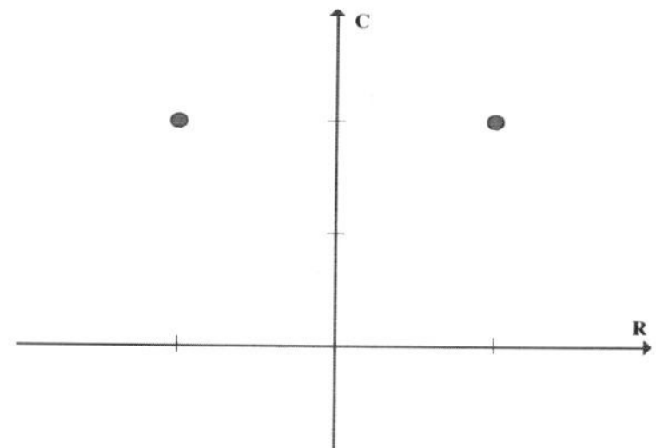

c. 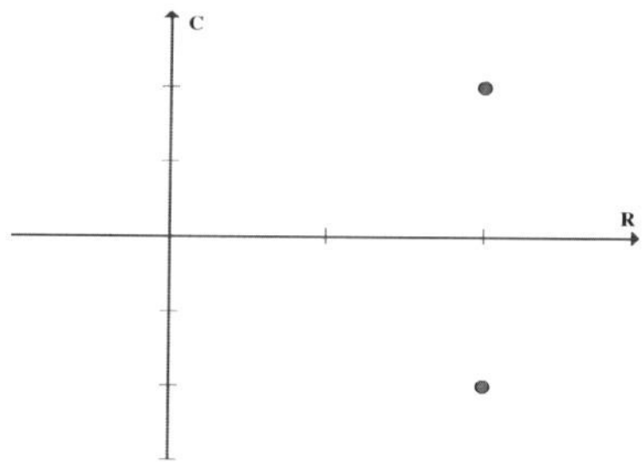

d. 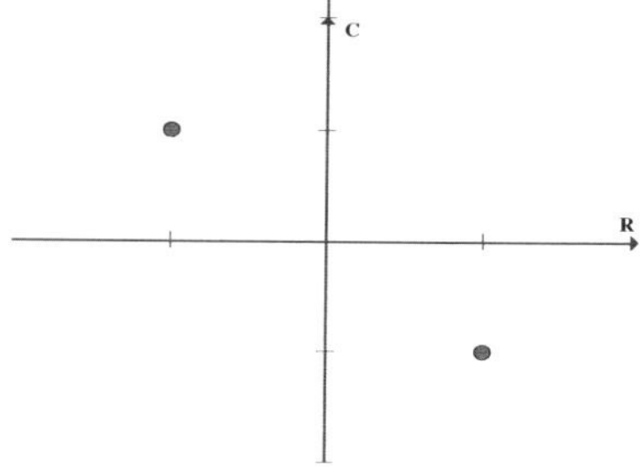

19. Find the midpoint between $(7 + 3i)$ and$(-3 - 5i)$ on the complex plane.

a. $10 + 8i$
b. $5 + 4i$
c. $4 - 2i$
d. $2 - i$

20. Which of the graphs correctly shows $(2 - 3i) - (1 + 4i)$ with the complex numbers drawn as vectors?

a. 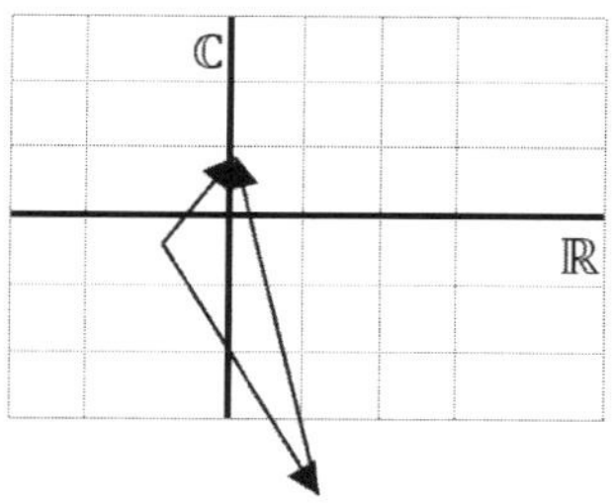

b. 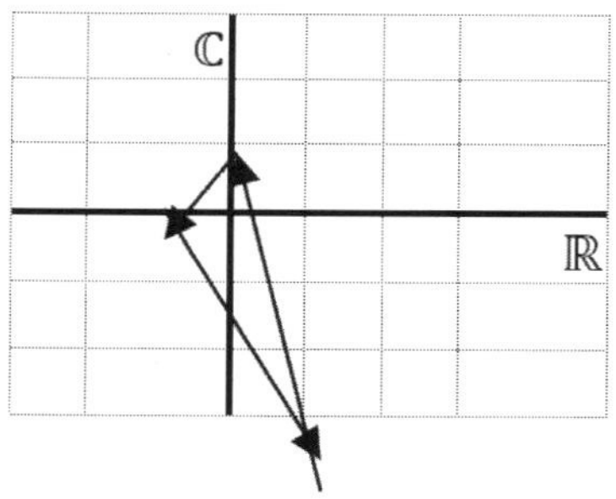

c. 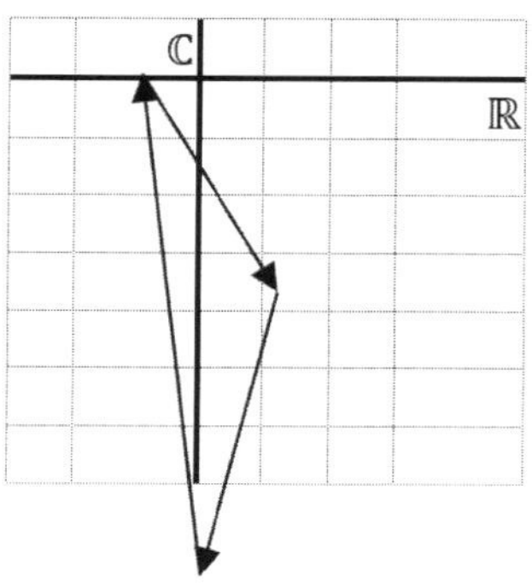

d. 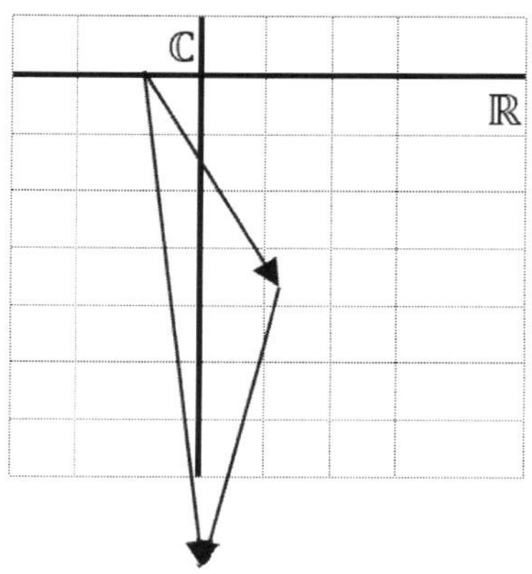

21. Rewrite in factored form: $x^2 - 6ix - 9$.

a. $i(x-3)^2$
b. $(x-3)(x-3i)$
c. $i(x-3)(x+3)$
d. $(x-3i)^2$

22. Find the roots: $y = 3x^2 - 7x + 11$.

a. $\frac{7 \pm i\sqrt{83}}{6}$
b. $\frac{7 \pm \sqrt{83}}{6}$
c. $\frac{7 \pm i\sqrt{181}}{6}$
d. $\frac{7 \pm \sqrt{181}}{6}$

23. Solve for x: $8x^2 = 12x - 30$.

a. $\frac{-12 \pm \sqrt{816}}{-16}$
b. $\frac{-3 \pm 4i\sqrt{51}}{-4}$
c. $\frac{3 \pm i\sqrt{51}}{4}$
d. $\frac{3 \pm \sqrt{51}}{4}$

24. Factor: $121x^2 + 98$.

a. $(11x - 49i)(11x + 2i)$
b. $(11x - 7i)(11x + 14i)$
c. $(11x - 14i)(11x - 7i)$
d. $(11x - 7i\sqrt{2})(11x + 7i\sqrt{2})$

25. Determine the number of roots in the polynomial: $y = ax^n + bx^{n-1} + cx^{n-2}$.

a. 3
b. a
c. n
d. n - 2

26. Determine the roots of $y = x^4 - 81$.

a. $9, -9$
b. $\pm 3, \pm 3i$
c. $-9, \pm 3$
d. $-9i, \pm 3$

27. Determine the magnitude and direction of the graphed vector:

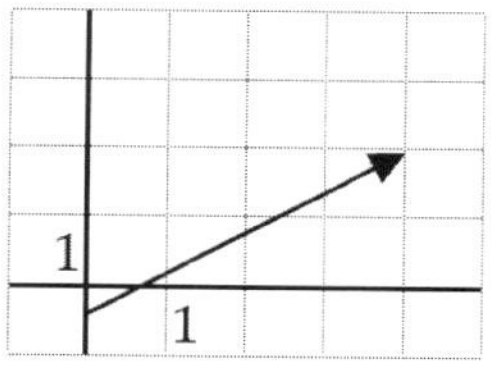

a. $2\sqrt{5}, 27°$
b. $2\sqrt{5}, 63°$
c. $20, 27°$
d. $20, 63°$

28. Find the magnitude and direction of vector $\overrightarrow{GH}$, from point G = (1,1) to point H = (13,6).

a. 13, 67°
b. 13, 23°
c. 12.2, 67°
d. 12.2, 23°

29. Which statement about vectors $\overrightarrow{MN}$ and $\overrightarrow{NM}$ is true?

a. $|\overrightarrow{MN}| = -|\overrightarrow{NM}|$
b. The directions of $\overrightarrow{MN}$ and $\overrightarrow{NM}$ have a difference of 180°.
c. $\overrightarrow{MN}$ and $\overrightarrow{NM}$ are the same in polar form.
d. All of the above.

30. Find the magnitude and direction of $\overrightarrow{AB}$ if A = (6,7) and B = (-2,-1).

a. $\sqrt{52}$, 56°
b. $\sqrt{52}$, 236°
c. $8\sqrt{2}$, 45°
d. $8\sqrt{2}$, 225°

31. A plane is traveling due west at a constant air speed of 350 mph. There are 50-mph winds blowing 30° west of north. Determine the actual speed and direction of the plane.

a. 377 mph at 7° north of west
b. 328 mph at 7° north of west
c. 394 mph at 3° north of west
d. 353 mph at 3° north of west

32. A boat is traveling 45° north of east on a river at a constant speed of 40 mph with respect to the water. The current flows at 15 mph 50° south of west. Determine the actual speed and direction of the boat.

a. 37 mph at 22° N of E
b. 37 mph at 50° N of E
c. 25 mph at 42° N of E
d. 25 mph at 53° N of E

33. Find the magnitude and direction of $\vec{m} + \vec{n}$ when $\vec{m}$ has magnitude 6 and direction 110° and $\vec{n}$ has magnitude 4 and direction 25°.

a. 7.5, 32°
b. 7.5, 78°
c. 7.2, 85°
d. 7.2, 95°

34. Define $\vec{s}$ to have a magnitude of 7 and a direction of 70° and $\vec{t}$ to have a magnitude of 5 and a direction of 140°. Find both the magnitude and direction of $\vec{t} - \vec{s}$.

a. 7, 210°
b. 7, 222°
c. 8.6, 210°
d. 8.6, 222°

35. Which of the following is not a correct depiction of $\vec{u} + \vec{v}$?

a.
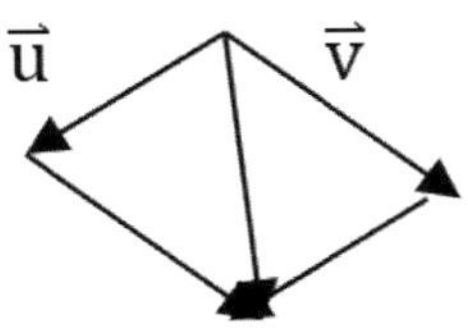

b.
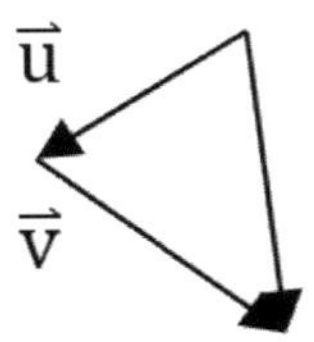

c.
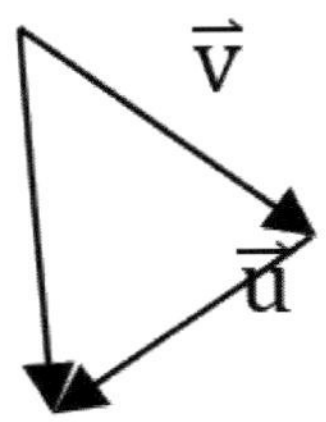

d.
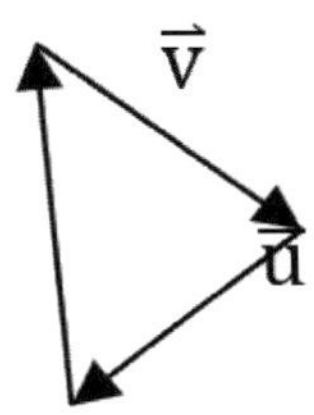

36. Solve for $\vec{a} - \vec{b}$, letting $\vec{a} = 5\hat{\imath} + 3\hat{\jmath}$ and $\vec{b} = 2\hat{\imath} - \hat{\jmath}$.

a. $3\hat{\imath} + 4\hat{\jmath}$
b. $3\hat{\imath} - 2\hat{\jmath}$
c. $7\hat{\imath} + 4\hat{\jmath}$
d. $3\hat{\imath} + 2\hat{\jmath}$

37. Write an expression for $\vec{u}$ in terms of $\vec{v}$.

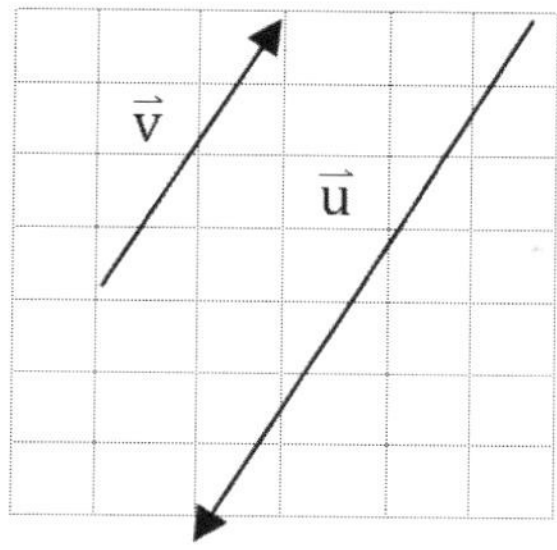

a. $\vec{u} = -\vec{v}$
b. $\vec{u} = 2\vec{v}$
c. $\vec{u} = -2\vec{v}$
d. $\vec{u} = 3\vec{v}$

38. Determine $5\vec{a}$ when $\vec{a} = (-3, 2)$.

a. (15, -10)
b. (-15, 10)
c. (-8, 3)
d. (2, 7)

39. Find the magnitude and direction of -4$\vec{u}$ when $\vec{u} = 2\hat{\imath} - \hat{\jmath}$.

a. $-4\sqrt{5}, -27°$
b. $4\sqrt{5}, -27°$
c. $-4\sqrt{5}, 153°$
d. $4\sqrt{5}, 153°$

40. The matrix represents the number of students in the class periods of each teacher. Which of the following statements can be made from the information in the matrix?

$$\begin{array}{c} \\ 1\text{st} \\ 2\text{nd} \\ 3\text{rd} \end{array} \begin{array}{c} \text{Smith} \quad \text{Tan} \\ \begin{bmatrix} 31 & 25 \\ 29 & 27 \\ 34 & 30 \end{bmatrix} \end{array}$$

a. Smith has 12 more students than Tan.
b. Smith is more popular with students than Tan.
c. Tan has smaller class sizes because she teaches harder classes.
d. Smith wants 3 students from 1st period to transfer into Tan's class.

41. The matrix represents the sales quotas in millions (of dollars) for each sales quarter (1 through 4) in the two divisions. What matrix represents a 10% increase in the quotas?

	West	East
1	1.5	2
2	1.7	2.1
3	1.7	2.1
4	1.9	2.3

a. $\begin{bmatrix} 1.51 & 2.01 \\ 1.71 & 2.11 \\ 1.71 & 2.11 \\ 1.91 & 2.31 \end{bmatrix}$

b. $\begin{bmatrix} 1.6 & 2.1 \\ 1.8 & 2.2 \\ 1.8 & 2.2 \\ 2.0 & 2.4 \end{bmatrix}$

c. $\begin{bmatrix} 1.65 & 2.2 \\ 1.87 & 2.31 \\ 1.87 & 2.31 \\ 2.09 & 2.53 \end{bmatrix}$

d. $\begin{bmatrix} 1.35 & 1.8 \\ 1.53 & 1.89 \\ 1.53 & 1.89 \\ 1.71 & 2.07 \end{bmatrix}$

42. Compute: $\begin{bmatrix} -7 & 4 \\ -2 & -3 \end{bmatrix} + \begin{bmatrix} -2 & 4 \\ -2 & 4 \end{bmatrix}$.

a. $\begin{bmatrix} -5 & 0 \\ 0 & 1 \end{bmatrix}$

b. $\begin{bmatrix} -5 & 8 \\ -4 & 7 \end{bmatrix}$

c. $\begin{bmatrix} -9 & 8 \\ -5 & 1 \end{bmatrix}$

d. $\begin{bmatrix} -9 & 8 \\ -4 & 7 \end{bmatrix}$

43. Compute: $\begin{bmatrix} 0 & 2 & -4 \\ 1 & 7 & 1 \end{bmatrix} - 3\begin{bmatrix} 5 & 3 & -1 \\ 5 & 0 & -4 \end{bmatrix}$.

a. $\begin{bmatrix} -5 & -1 & -3 \\ -4 & 7 & 5 \end{bmatrix}$

b. $\begin{bmatrix} -3 & -1 & 0 \\ -2 & 4 & 2 \end{bmatrix}$

c. $\begin{bmatrix} -15 & -7 & -1 \\ -14 & 7 & 13 \end{bmatrix}$

d. $\begin{bmatrix} -15 & -11 & -7 \\ -14 & 7 & -11 \end{bmatrix}$

44. For $A \cdot A^{-1} = I$, which of the following statements must be true?

a. A must be a square matrix.
b. detA $\neq 0$
c. A·I=A
d. All of the above.

45. Describe how the matrix transforms the vector: $\begin{bmatrix} -3 & 0 \\ 0 & 3 \end{bmatrix} \cdot \begin{bmatrix} -1 \\ -2 \end{bmatrix}$.

a. Reflected across the x-axis and tripled in length.
b. Reflected across the y-axis and tripled in length.
c. Reflected across the origin and scaled by a factor of 3.
d. Rotated 90° and scaled by a factor of 3.

46. Evaluate: $\begin{bmatrix} 1 & -2 & 3 \\ 5 & -2 & -1 \end{bmatrix} \cdot \begin{bmatrix} 6 & -3 \\ 4 & 5 \\ 2 & 8 \end{bmatrix}$.

a. $\begin{bmatrix} 4 & 11 \\ 20 & -33 \end{bmatrix}$
b. $\begin{bmatrix} 14 & 12 \\ 14 & 12 \end{bmatrix}$
c. $\begin{bmatrix} -8 & -31 \\ 20 & -3 \end{bmatrix}$
d. $\begin{bmatrix} 20 & 37 \\ 40 & 40 \end{bmatrix}$

47. Create a matrix that will reflect $\vec{m}$ over the origin and double the magnitude, if $\vec{m} = \begin{bmatrix} 3 \\ 2 \\ -1 \end{bmatrix}$.

a. $\begin{bmatrix} 2 & 2 & 2 \\ 2 & 2 & 2 \\ 2 & 2 & 2 \end{bmatrix}$
b. $\begin{bmatrix} -2 & 2 & 2 \\ 2 & -2 & 2 \\ 2 & 2 & -2 \end{bmatrix}$
c. $\begin{bmatrix} 0 & 0 & -2 \\ 0 & -2 & 0 \\ -2 & 0 & 0 \end{bmatrix}$
d. $\begin{bmatrix} -2 & 0 & 0 \\ 0 & -2 & 0 \\ 0 & 0 & -2 \end{bmatrix}$

48. Find the determinant of M, and determine if M^{-1} exists, if $M = \begin{bmatrix} 25 & -10 \\ 15 & -6 \end{bmatrix}$.

a. $\det M = 0, M^{-1}$ exists
b. $\det M = 0, M^{-1}$ does not exist
c. $\det M = 300, M^{-1}$ exists
d. $\det M = 300, M^{-1}$ does not exist

49. Determine the area of the parallelogram created by $\vec{m} = \begin{bmatrix} 5 \\ 7 \end{bmatrix}$ and $\vec{n} = \begin{bmatrix} -3 \\ 10 \end{bmatrix}$.

a. 31
b. 55
c. 71
d. 85

50. Which of the following properties does *not* hold true for the multiplication of square matrices?

a. Associative Property
b. Commutative Property
c. Distributive Property
d. None of the above; the above properties all hold true.

Answers and Explanations

Math.Content.N.RN.2

1. B: First, simplify within the parentheses by adding the exponents since the bases are equal: $4^3 \cdot 4^3 = 4^{3+3} = 4^6$. Next convert the radical to a fractional exponent: $\sqrt[5]{(4^6)^2} = (4^6)^{2/5}$. Finally, multiply the exponents: $(4^6)^{2/5} = 4^{12/5}$.

Math.Content.N.RN.3

2. D: An irrational numbers is a number that cannot be written as a fraction. Some examples of irrational numbers are π, $\sqrt{3}$, and e. If an irrational number is added to or subtracted from, it will still not be a repeating decimal or expressible as a fraction. (The restriction in choice C prevents the subtraction from equaling zero, which would be a rational exception to the rule.) On the other hand, while most irrational products give irrational answers, there are counterexamples: $\sqrt[5]{21} \cdot \sqrt[5]{21^4} = 21^{1/5} \cdot 21^{4/5} = 21$. Since at least one example exists, D is false.

Math.Content.N.RN.1

3. C: Start by rewriting the problem using fractional exponents. Remember that fractional exponential notation is $\sqrt[y]{b^x} = b^{x/y}$. Simplify when the bases are the same, the general rules for which are $b^x \cdot b^y = b^{x+y}$ and $\frac{b^x}{b^y} = b^{x-y}$:

$$\frac{3 \cdot 3 \cdot \sqrt[3]{3 \cdot 3} \cdot 5 \cdot 5 \cdot \sqrt[3]{7 \cdot 7 \cdot 7 \cdot 7}}{3 \cdot \sqrt[3]{5 \cdot 5} \cdot 7\sqrt[3]{7 \cdot 7}} = \frac{3^2 \cdot 3^{2/3} \cdot 5^2 \cdot 7^{4/3}}{3^1 \cdot 5^{2/3} \cdot 7^1 \cdot 7^{2/3}} = 3^{(2+2/3-1)} \cdot 5^{(2-2/3)} \cdot 7^{(4/3-1-2/3)}$$

$$= 3^{5/3} \cdot 5^{4/3} \cdot 7^{-1/3}$$

Math.Content.N.RN.2

4. A: Since the bases are the same, start by combining the exponents in the parentheses using the rule $b^x \cdot b^y = b^{x+y}$. Then multiply the exponent outside the parentheses to the exponent inside the parentheses using the rule $(b^x)^y = b^{x \cdot y}$:

$$(7^{12} \cdot 7^{-5})^{2/3} = (7^{12-5})^{2/3} = (7^7)^{2/3} = 7^{7 \cdot 2/3} = 7^{14/3}$$

Math.Content.N.RN.3

5. D: One way to answer this question is calculate the answers and look for the number that can be written as a fraction. $5\sqrt[3]{49} + 1 = 19.2965$, and therefore $\sqrt[3]{49}$ is irrational. $\sqrt{99} \cdot \sqrt{7} + 3 = 29.3248$, since both $\sqrt{99}$ and $\sqrt{7}$ are irrational. 3π-4=5.4247 since π is irrational. $2\sqrt{169}$=26 is, of course, rational.

Math.Content.N.Q.2

6. A: Set this problem up using dimensional analysis. Convert the various statements into fractions that express equivalence; for example, write "90 mg per kg of body weight per day" as $\frac{90\frac{\text{mg}}{\text{kg}}}{\text{day}} = \frac{90mg}{kg} \cdot \frac{1}{day} = \frac{90mg}{kg \cdot day}$. Multiply the fractions and/or their reciprocals together so that unwanted units will cancel, leaving only the desired unit at the end:

$45\ lbs \cdot \frac{1\text{ kg}}{2.2\text{ lbs}} \cdot \frac{90\text{ mg}}{\text{kg}\cdot\text{day}} \cdot 10\ days \cdot \frac{1\ gram}{1000\ mg} = 45\ \cancel{lbs} \cdot \frac{1\ \cancel{\text{kg}}}{2.2\ \cancel{\text{lbs}}} \cdot \frac{90\ \cancel{\text{mg}}}{1\ \cancel{\text{kg}} \cdot 1\ \cancel{\text{day}}} \cdot 10\ \cancel{days} \cdot \frac{1\ gram}{1000\ \cancel{mg}} = \frac{45\cdot 90\cdot 10}{2.2\cdot 1000} grams = 18.4\ grams.$

Math.Content.N.Q.1

7. B: Set up conversion fractions to find the total cost of gas for the trip, and multiply the factors so unwanted units cancel (leaving dollars): $(77\ miles)(2\ ways) \cdot \frac{1\ gallon}{23\ miles} \cdot \frac{\$3.89}{gallon} = \$26.05$. If they split the cost evenly, each pays a third of $26.05, or about $8.68. (The size of the gas tank is irrelevant!)

Math.Content.N.Q.1

8. C: Heidi's recipe calls for 1 tablespoon for 2 loaves, so for 350 loaves she would need 175 tablespoons. There are 3 teaspoons in each tablespoon, though, so asking for 525 teaspoons seems silly. There are 16 tablespoons in a cup, so she would need $175\ tbs \cdot \left(\frac{1\ cup}{16\ tbs}\right) = 10.93$ cups. Similarly, there are 2 cups in a pint so she would need 5.46 pints, and there are 2 pints in a quart so she would need 2.73 quarts. There are 4 quarts in a gallon, though, so she would need 0.683 gallons, which feels a little awkward. Quarts is the best unit here.

Math.Content.N.Q.3

9. D: First, determine the range of values for each axis. Even though n represents the number of years since 1986-7, the time period starts in 1987 and runs until 2012, so the horizontal (time) axis should be labeled as in choices B, C, or D. To find the maximum and minimum number of students per computer, evaluate the model at $n = 1$ and 25: in 1987, $a_1 = 35.8127(0.8646)^1 = 30.96 \approx 31$ students per computer; in 2012, n=2012-1987=25, and $a_{25} = 35.8127(0.8646)^{25} = .9428 \approx$ 1 student per computer. The y-axis, then, needs to show a range of students per computer [1, 31]. The range of the y-axis shown in choice C is larger than it needs to be, and while not technically incorrect, this makes choice D the *most* appropriate choice of scale.

Math.Content.N.Q.3

10. D: The Earth revolves around the sun once every year, or 365 days. Converting this to seconds: $365\ days \cdot \left(\frac{24\ hours}{1\ day}\right) \cdot \left(\frac{60\ minutes}{1\ hour}\right) \cdot \left(\frac{60\ seconds}{1\ minute}\right) = 31{,}536{,}000\ seconds$. Evaluating the expression for this value of t: $[(3.35 \times 10^{18})(31{,}536{,}000)^2]^{\frac{1}{3}} = 1.49 \times 10^{11}$. This answer is in meters. Reporting this answer, one should use the largest possible unit; gigameters is the largest unit shown ($1\ Gm = 10^9\ m$).

Math.Content.N.CN.1

11. A: The order of operations is Parentheses, Exponent, Multiplication/Division, Addition/Subtraction; this is often represented by the acronym PEMDAS. Radicals are not listed explicitly, because they can be written as parentheses with an exponent: $\sqrt[m]{x+y} = (x+y)^{1/m}$. The square root of a negative number is imaginary and can be written in terms of i: $\sqrt{-1} = i$. Therefore:

$$3\sqrt{4-5\cdot 8} + \sqrt{5-6} = 3(4-5\cdot 8)^{½} + (5-6)^{½} = 3(-36)^{½} + (-1)^{½} = 3\cdot 6i + i = 18i + i = 19i$$

Math.Content.N.CN.2

12. C: Start by distributing each coefficient across the parentheses. Next add the complex binomials. Complex numbers have a real part, *a,* and an imaginary part, *b*, and are usually written as *a* + *bi*. When adding imaginary numbers, add the real parts and the imaginary parts separately:

$$4(-3+5i)+6(1-3i)=-12+20i+6-18i=(-12+6)+(20i-18i)=-6+2i$$

Math.Content.*N.CN.2*

13. A: Multiplication of complex numbers is very similar in process to multiplying binomials. Use the FOIL method: (a+bi)(c+di)=$\overset{\text{First}}{a\cdot c}\ \overset{\text{Outer}}{+a\cdot di}\ \overset{\text{Inner}}{+bi\cdot c}\ \overset{\text{Last}}{+bi\cdot di}$. The last term will have an i^2, which needs to be simplified using $i^2=\sqrt{(-1)^2}=-1$. Last, combine the real terms and the imaginary terms:

$$\begin{aligned}(9-3i)(-7-12i) &= (9)(-7)+(9)(-12i)+(-3i)(-7)+(-3i)(-12i)\\ &= -63-108i+21i+36i^2=-63-108i+21i-36=-99-87i\end{aligned}$$

Math.Content.N.CN.3

14. B: For every complex number, $z=a+bi$, there is a conjugate found by replacing *i* with –*i*, $\bar{z}=a-bi$. In this problem, $z=8-6i$, so $\bar{z}=8+6i$.

Math.Content.N.CN.3

15. D: Standard form for complex numbers requires the denominator be completely real, that is, have no imaginary part. When dividing complex numbers, $\frac{a+bi}{c+di}$, first, multiply by "a well-chosen 1"; that is, multiply the quotient by a fraction with the value 1 written so as the product of the denominators is completely real. This is accomplished by multiplying by the conjugate of the denominator over itself: $\frac{c-di}{c-di}$. Then, FOIL both numerator and denominator, and combine like terms and simplify, remembering that $i^2=-1$:

$$\frac{15-5i}{4-2i}\cdot\frac{4+2i}{4+2i}=\frac{60+30i-20i-10i^2}{16+8i-8i-4i^2}=\frac{60+10i+10}{16+4}=\frac{70+10i}{20}=\frac{10(7+i)}{2\cdot 10}=\frac{7+i}{2}$$

Math.Content.N.CN.4

16. B: First, determine the complex number in rectangular form. Rectangular form is $a+bi$, graphed with a on the real (horizontal) axis and b on the complex (vertical) axis: $1+4i$. Next, convert rectangular form to polar form: $r(cos\theta+isin\theta)$, with $r=\sqrt{a^2+b^2}$ and $\theta=\tan^{-1}\frac{b}{a}$. (If the complex number were in quadrants 2 or 3, $\theta=\tan^{-1}\frac{b}{a}$ yields a reference angle rather than the actual direction.) For $1+4i$, $r=\sqrt{1^2+4^2}=\sqrt{1+16}=\sqrt{17}$. The complex number is in quadrant 1, so $\theta=\tan^{-1}\frac{4}{1}\approx 76°$. $1+4i=\sqrt{17}(\cos 76°+i\sin 76°)$.

Math.Content.N.CN.5

17. C: The distance formula, $D=\sqrt{(x_1-x_2)^2+(y_1-y_2)^2}$, applied to the complex plane, becomes $D=\sqrt{(a_1-a_2)^2+(b_1-b_2)^2}$. Substitute the coordinates and solve: $D=\sqrt{(3-(-2))^2+((-6)-10)^2}=\sqrt{5^2+(-16)^2}=\sqrt{281}=16.76$.

Math.Content.N.CN.6

18. C: A conjugate pair is a complex number, $z = a + bi$, and its conjugate, $\overline{z} = a - bi$. Graphed on the complex plane, since the vertical coordinate changes sign while the horizontal component is unchanged, $\overline{z}$ is z reflected across the real axis. Graph C fits this description.

Math.Content.N.CN.6

19. D: Find the midpoint by adapting the expressions for midpoint in the normal Cartesian plane, $M = \left(\frac{x_1+x_2}{2}, \frac{y_1+y_2}{2}\right)$, to complex numbers: $M = \left(\frac{a_1+a_2}{2}, \frac{b_1+b_2}{2}\right)$. Substitute the coordinates to obtain: $M = \left(\frac{7-3}{2}, \frac{3-5}{2}\right) = \left(\frac{4}{2}, -\frac{2}{2}\right) = (2, -1)$, and the midpoint is 2 - *i*.

Math.Content.N.CN.5

20. D: Vector subtraction can be thought of the vector addition of the opposite vector, which in this case yields $(2 - 3i) + (-1 - 4i)$. $(2 - 3i)$ is a vector in the 4th quadrant; the vector $(-1 - 4i)$ points to the 3rd quadrant. Position the tail of the latter at the tip of the former, and the resultant is drawn from the tail of the former to the tip of the latter. In both C and D the individual vectors are graphed correctly, but only in choice D is the resultant correctly drawn from the tail of the first vector to the tip of the second.

Math.Content.N.CN.8

21. D: A common factoring pattern is $(x + a)^2 = x^2 + 2ax + a^2$. Looking at our problem, to factor $x^2 - 6ix - 9$, the i requires complex numbers and the 6 and 9 remind of the above pattern. Therefore, $\mathrm{a}^2 = -9$ and $2\mathrm{ax} = -6\mathrm{ix}$, so $a = -3i$, and the trinomial factors to $(x - 3i)^2$. To check: $(x - 3i)^2 = x^2 - 3ix - 3ix + 9i^2 = x^2 - 6ix - 9$.

Math.Content.N.CN.7

22. A: To find the roots, substitute the coefficients into the quadratic formula. When $y = ax^2 + bx + c$, $x = \frac{-b\pm\sqrt{b^2-4ac}}{2a}$. In this case:

$$x = \frac{7 \pm \sqrt{(-7)^2 - 4 \cdot 3 \cdot 11}}{2 \cdot 3} = \frac{7 \pm \sqrt{49 - 132}}{6} = \frac{7 \pm \sqrt{-83}}{6} = \frac{7 \pm \mathrm{i}\sqrt{83}}{6}$$

Math.Content.N.CN.7

23. C: To solve for x, subtract $12x - 30$ from both sides of the equation: $8x^2 - 12x + 30 = 0$. Now the equation is in standard form, $a\mathrm{x}^2 + \mathrm{bx} + \mathrm{c} = 0$, and the coefficients can be used in the quadratic formula to solve for x:

$$x = \frac{-b \pm \sqrt{b^2 - 4ac}}{2a} = \frac{12 \pm \sqrt{(-12)^2 - 4(8)(30)}}{2(8)} = \frac{12 \pm \sqrt{144 - 960}}{16} = \frac{12 \pm \sqrt{-816}}{16}$$
$$= \frac{12 \pm 4\mathrm{i}\sqrt{51}}{16} = \frac{4(3 \pm i\sqrt{51})}{4(4)} = \frac{3 \pm i\sqrt{51}}{4}$$

Math.Content.N.CN.8

24. D: The difference of squares is a very common factoring pattern, $x^2 - y^2 = (x - y)(x + y)$. This problem requires factoring a sum of squares. When complex numbers are taken into account, the sum of squares can be written as a difference of squares: $x^2 + y^2 = x^2 - (iy)^2$. With that substitution:

$$121x^2 + 98 = (11x)^2 - \left(i\sqrt{98}\right)^2 = \left(11x - 7i\sqrt{2}\right)\left(11x + 7i\sqrt{2}\right)$$

Math.Content.N.CN.9

25. C: The Fundamental Theorem of Algebra states that the degree of a polynomial with complex coefficients is equal to the number of complex roots for that polynomial (including multiple instances of the same root). Note, too, that real coefficients and roots are included in the set of complex numbers and simply have imaginary parts equal to zero. The degree of a polynomial is the highest power of the independent variable. In the given equation $y = ax^n + bx^{n-1} + cx^{n-2}$, n is the highest power of the independent variable, x. So, n is the degree and the number of roots of this polynomial.

Math.Content.N.CN.9

26. B: As written this equation is a Difference of Squares: $y = x^4 - 81 = (x^2 - 9)(x^2 + 9)$. The factors, though, are themselves the difference (and sum) of squares! Factor the difference of squares as normal, and the sum of squares using $x^2 + y^2 = x^2 - (iy)^2$ to obtain $(x^2 - 9)(x^2 + 9) = (x - 3)(x + 3)(x - 3i)(x + 3i)$. A root, or zero, of the function occurs when $y = 0$, or when the graph crosses the x-axis. To find the roots, set the factored form equal to zero. Since a product is only equal to zero if one of the factors is zero, set each factor equal to zero to obtain x = {3,-3, 3i, -3i}.

Math.Content.N.VM.1

27. A: The magnitude of the vector is the length of the vector and can be found using the Pythagorean Theorem. The vertical and horizontal components can be measured directly on the graph. Magnitude = $\sqrt{2^2 + 4^2} = \sqrt{20} = 2\sqrt{5}$. The direction of the vector is the measure of the angle between the vector and the positive side of the x-axis. In quadrants I and IV, the direction can be found using the tangent function. (In quadrants II and III, the tangent function yields a reference angle only.) The vector is in quadrant I, so θ is the direction: $\theta = \tan^{-1}\frac{y}{x} = \tan^{-1}\frac{2}{4} = 26.56 \approx 27°$.

Math.Content.N.VM.2

28. B: Find the magnitude using the distance formula: $\left|\overrightarrow{GH}\right| = \sqrt{(x_1 - x_2)^2 + (y_1 - y_2)^2} = \sqrt{(1 - 13)^2 + (1 - 6)^2} = \sqrt{(-12)^2 + (-5)^2} = \sqrt{144 + 25} = 13$. The vector points into quadrant I, so its direction can be found with the tangent function: $\theta = \tan^{-1}\left(\frac{y_1 - y_2}{x_1 - x_2}\right) = \tan^{-1}\left(\frac{1-6}{1-13}\right) = \tan^{-1}\left(\frac{-5}{-12}\right) = 22.61 \approx 23°$.

Math.Content.N.VM.1

29. B: $\overrightarrow{MN}$ and $\overrightarrow{NM}$ have the same endpoints and opposite directions. The magnitude of both vectors is the same, $\left|\overrightarrow{MN}\right| = \left|\overrightarrow{NM}\right|$, because the length between points M and N is always the same.

The vectors are in opposite directions, so the directions have a difference of 180°. (This difference in direction will be clear in polar form. If $\overrightarrow{MN} = r(cos\theta + isin\theta)$, then $\overrightarrow{NM} = r\big(cos(\theta + 180) + isin(\theta + 180)\big)$.)

Math.Content.N.VM.2

30. D: Find the magnitude using the distance formula: $|\overrightarrow{AB}| = \sqrt{(x_1 - x_2)^2 + (y_1 - y_2)^2} = \sqrt{\big(6 - (-2)\big)^2 + \big(7 - (-1)\big)^2} = \sqrt{8^2 + 8^2} = \sqrt{2(8^2)} = 8\sqrt{2}$. To find the direction, we need to know the quadrant to which the vector points.

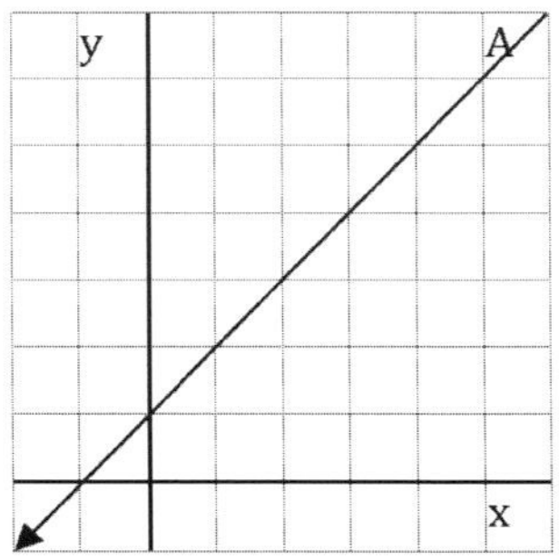

If the initial point were at the origin, the vector would (still) be pointing into the third quadrant. θ is our reference angle. In the third quadrant, the direction is 180+θ. Find the direction using the inverse tangent function: $\theta = \tan^{-1}\left(\frac{y_1 - y_2}{x_1 - x_2}\right) = \tan^{-1}\left(\frac{7-(-1)}{6-(-2)}\right) = \tan^{-1}\left(\frac{8}{8}\right) = 45°$, and the direction is 180 + 45=225°.

Math.Content.N.VM.3

31. A: Using the information in the problem, sketch the vector addition triangle. Determine and label as many angles as possible. Then draw the triangle with only the angles and magnitudes.

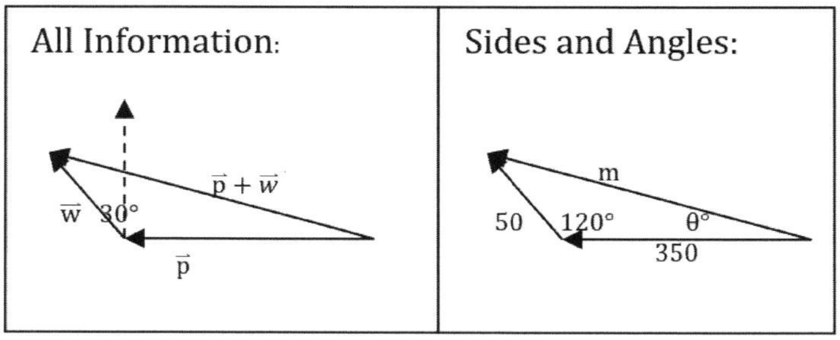

Use the Law of Cosines to find m:

$$m = \sqrt{a^2 + b^2 - 2 \cdot a \cdot b \cdot cos\theta} = \sqrt{350^2 + 50^2 - 2 \cdot 350 \cdot 50 \cos 120} \approx 377$$

Use the Law of Sines to find θ: $\frac{sin120}{377} = \frac{sin\theta}{50}$. Multiply by 50 on both sides to obtain: $\sin\theta = \frac{50 \cdot \sin 120}{377}$ $\xrightarrow{yields} \theta = \sin^{-1}(.11486) = 6.59 \approx 7$ The actual speed, then, is 377 and the actual direction is 7° north of west.

Alternatively, write the vectors representing the plane's and wind's velocities using unit vectors, and add them: $\vec{p} = -350\hat{\imath} + 0\hat{\jmath}$, and $\vec{w} = (50\cos 120)\hat{\imath} + (50\sin 120)\hat{\jmath} = -25\hat{\imath} + (25\sqrt{3})\hat{\jmath}$, so that $\vec{p} + \vec{w} = (-350 + (-25))\hat{\imath} + (0 + 25\sqrt{3})\hat{\jmath} = -375\hat{\imath} + (25\sqrt{3})\hat{\jmath}$. Find the magnitude using the distance formula and the reference angle: $|\vec{p} + \vec{w}| = \sqrt{(-375)^2 + (25\sqrt{3})^2} \approx 377$, and $180° + \theta = 180° + \tan^{-1}\frac{25\sqrt{3}}{-375} \approx 180° + (-7°)$, which is 7° north of west.

Math.Content.N.VM.3

32. C: Sketch a diagram of the information in the problem. Label as many angles as possible.

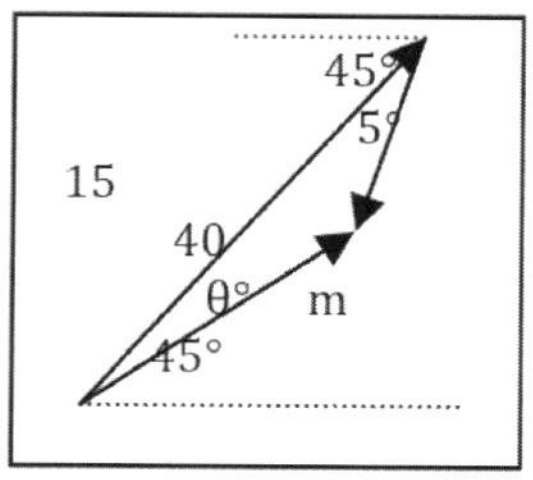

Use the Law of Cosines to solve for the missing magnitude:

$$m = \sqrt{40^2 + 15^2 - 2 \cdot 40 \cdot 15 \cdot \cos 5} = 25$$

Use the Law of Sines to find θ. (Keep in mind that θ is not the direction of the boat but an angle in the triangle representing the problem.)

$$\frac{sin5}{25} = \frac{sin\theta}{15} \xrightarrow{yields} \theta = \sin^{-1}\left(\frac{15 \cdot sin5}{25}\right) \approx 3°$$

The actual direction of the boat is found by 45 – θ = 42°.

Alternatively, write the vectors representing the boat's and current's velocities using unit vectors, and add them: $\vec{b} = (40\cos 45)\hat{\imath} + (40\sin 45)\hat{\jmath}$, and $\vec{c} = (15\cos 230)\hat{\imath} + (15\sin 230)\hat{\jmath}$, so that $\vec{b} + \vec{c} = 18.64\hat{\imath} + 16.79\hat{\jmath}$. Find the magnitude using the distance formula and the reference angle: $|\vec{b} + \vec{c}| = \sqrt{18.64^2 + 16.79^2} \approx 25$, and $\theta = \tan^{-1}\frac{16.79}{18.64} \approx 42°$.

Math.Content.N.VM.4b

33. B: Draw out the vector addition. Determine and label as many angles as possible. Draw the triangle again with only magnitudes and interior angles.

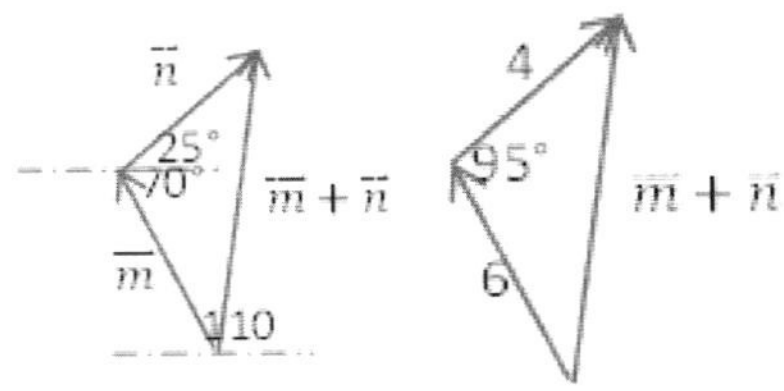

Use the Law of Cosines to determine $|\vec{m}+\vec{n}| = \sqrt{4^2+6^2-2\cdot4\cdot6\cdot\cos 95} = 7.5$. Use Law of Sines to find θ: $\frac{sin95}{7.5} = \frac{sin\theta}{4} \xrightarrow{yields} \theta = \sin^{-1}\left(\frac{4\cdot\sin 95}{7.5}\right) = 32°$. θ is the angle in the triangle, but the direction is the angle from the 0° x-axis to the vector, so the direction is 110° – θ = 110 – 32 = 78°.

Math.Content.N.VM.4c

34. A: Draw out the problem with all the given information. Refine the sketch with only the essential information.

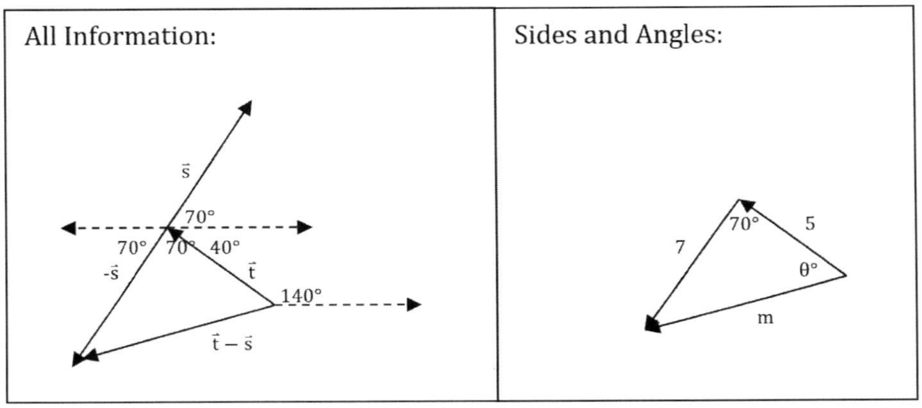

Use Law of Cosines to determine m, the magnitude of $\vec{t}-\vec{s}$: $m = \sqrt{7^2+5^2-2\cdot7\cdot5\cos 70} = 7$. Next use Law of Sines to find θ: $\frac{sin\theta}{7} = \frac{sin70}{7} \xrightarrow{yields} \theta = 70°$. The direction of $\vec{t}-\vec{s}$, then, is 140 + θ = 140 + 70 = 210°.

Math.Content.N.VM.4a

35. D: Vector addition can be done end-to-end, also called tip-to-tail, or with the parallelogram rule. End-to-end vector addition is done by drawing the second vector's initial point at the first vector's endpoint. Then the sum is drawn from the first vector's initial point to the second vector's endpoint. In this case, if $\vec{u}$ is from point A to point B and $\vec{v}$ is from point B to point C, then $\vec{u}+\vec{v}$ is from point A to point C. Vector addition is commutative so both B and C are correct depictions. Choice A shows the parallelogram rule; a parallelogram is created with the vectors and the sum is the diagonal from the initial point to the endpoint. Choice D, though, is end-to-end vector addition with the resultant pointing the wrong direction.

Math.Content.N.VM.4c

36. A: First, substitute the component values into the problem. Then distribute the negative, and combine like terms: $\vec{a} - \vec{b} = (5\hat{\imath} + 3\hat{\jmath}) - (2\hat{\imath} - \hat{\jmath}) = 5\hat{\imath} + 3\hat{\jmath} - 2\hat{\imath} + \hat{\jmath} = 3\hat{\imath} + 4\hat{\jmath}$.

Math.Content.N.VM.5a

37. C: There are multiple approaches to this problem. To proceed algebraically, write the component expressions for each vector: $\vec{v} = 2\hat{\imath} + 3\hat{\jmath}, \vec{u} = -4\hat{\imath} - 6\hat{\jmath}$. Then, factor: $\vec{u} = -2(2\hat{\imath} + 3\hat{\jmath}) = -2\vec{v}$. To proceed graphically, or visually, recognize that $\vec{u}$ is antiparallel to, and twice as long as, $\vec{v}$. Multiplying a vector by a scalar stretches the vector by that scalar factor, and multiplying by -1 reverses the direction, so C must be the answer.

Math.Content.N.VM.5a

38. B: To find $5\vec{a}$, remember that scalar multiplication of a vector simply requires multiplying each of the components by that scalar: $5\vec{a} = (5 \cdot -3, 5 \cdot 2) = (-15, 10)$.

Math.Content.N.VM.5b

39. D: First find $-4\vec{u}$ by distributing the -4 to the components: $-4\vec{u} = -4(2\hat{\imath} - \hat{\jmath}) = -8\hat{\imath} + 4\hat{\jmath}$. Then use the Pythagorean Theorem to find the magnitude: $|-4\vec{u}| = \sqrt{(-8)^2 + 4^2} = \sqrt{80} = 4\sqrt{5}$. Since (-8,4) is in quadrant II, -$4\vec{u}$ is also in quadrant II. Find the reference angle with the tangent function: $\theta = tan^{-1}\frac{4}{-8} = -26.56 \approx -27°$, and in quadrant II, the direction is 180 + θ = 180 – 27 = 153°.

Math.Content.N.VM.6

40. A: Statements B, C, and D all infer non-quantitative issues that are not presented in the matrix. Smith does have more students than Tan, but since the reasons this is the case are not presented in the matrix, we cannot conclude anything about those reasons.

Math.Content.N.VM.7

41. C: To show a 10% increase in sales, multiply each number by 1.1:

$$1.1\begin{bmatrix}1.5 & 2\\1.7 & 2.1\\1.7 & 2.1\\1.9 & 2.3\end{bmatrix} = \begin{bmatrix}1.1\cdot1.5 & 1.1\cdot2\\1.1\cdot1.7 & 1.1\cdot2.1\\1.1\cdot1.7 & 1.1\cdot2.1\\1.1\cdot1.9 & 1.1\cdot2.3\end{bmatrix} = \begin{bmatrix}1.65 & 2.2\\1.87 & 2.31\\1.87 & 2.31\\2.09 & 2.53\end{bmatrix}$$

Math.Content.N.VM.8

42. C: First, make sure the matrices have the same dimensions. Since they do, then, add the elements in corresponding positions and place that sum in that position in the resulting matrix: in other words,$a_{1,1} + b_{1,1} = (a + b)_{1,1}$. Repeat for each position in the matrix:

$$\begin{bmatrix}-7 & 4\\-2 & -3\end{bmatrix} + \begin{bmatrix}-2 & 4\\-2 & 4\end{bmatrix} = \begin{bmatrix}-7-2 & 4+4\\-2-2 & -3+4\end{bmatrix} = \begin{bmatrix}-9 & 8\\-4 & 1\end{bmatrix}$$

Math.Content.N.VM.8

43. C: First, make sure the matrices have the same dimensions as matrix addition and subtraction are only possible when the dimensions are the same. Then, complete the scalar multiplication by multiplying each element of the second matrix by 3. Finally, subtract the elements in the same positions:

$$\begin{bmatrix} 0 & 2 & -4 \\ 1 & 7 & 1 \end{bmatrix} - 3\begin{bmatrix} 5 & 3 & -1 \\ 5 & 0 & -4 \end{bmatrix} = \begin{bmatrix} 0 & 2 & -4 \\ 1 & 7 & 1 \end{bmatrix} - \begin{bmatrix} 3 \cdot 5 & 3 \cdot 3 & 3 \cdot -1 \\ 3 \cdot 5 & 3 \cdot 0 & 3 \cdot -4 \end{bmatrix} =$$

$$\begin{bmatrix} 0 & 2 & -4 \\ 1 & 7 & 1 \end{bmatrix} - \begin{bmatrix} 15 & 9 & -3 \\ 15 & 0 & -12 \end{bmatrix} = \begin{bmatrix} 0-15 & 2-9 & -4+3 \\ 1-15 & 7-0 & 1+12 \end{bmatrix} = \begin{bmatrix} -15 & -7 & -1 \\ -14 & 7 & 13 \end{bmatrix}$$

Math.Content.N.VM.10

44. D: The question asks what must be true for $\boldsymbol{A} \cdot \boldsymbol{A}^{-1} = \boldsymbol{I}$. In words, under what conditions will matrix A have an inverse? To have an inverse, a matrix must have the same number of rows as columns; it must be square (statement A). The determinant of a square matrix A must be nonzero for the inverse to exist (statement B). In order for the product of matrix A and it's inverse to equal I, we must define the identity matrix, I (statement C). All three conditions are required; D is correct.

Math.Content.N.VM.11

45. B: The ±3's on the diagonal indicate the matrix will triple the length of the vector. The negative in position (1,1) will cause a reflection across the y-axis (from quadrant 3 to quadrant 4): $\begin{bmatrix} -3 & 0 \\ 0 & 3 \end{bmatrix} \cdot \begin{bmatrix} -1 \\ -2 \end{bmatrix} = \begin{bmatrix} -3 \cdot -1 + 0 \cdot -2 \\ 0 \cdot -1 + 3 \cdot -2 \end{bmatrix} = \begin{bmatrix} 3 \\ -6 \end{bmatrix}$.

Math.Content.N.VM.8

46. A: First, check that matrix multiplication is possible: matrix multiplication is only possible when the number of columns in matrix A is equal to the number of rows in matrix B. Since the two matrices are of dimension (2 x 3) · (3 x 2), matrix multiplication is possible (the inner pair of dimensions are equal) and the dimension of the result is 2 x 2 (the outer pair). To find an element of the product, multiply each entry of the corresponding row of matrix A by the respective entry (in order) of the corresponding column of matrix B and add the products. As a formula, $c_{11} = a_{11} \cdot b_{11} + a_{12} \cdot b_{21} + a_{13}b_{31}$. (The position of each element is given by the subscripts: first the row number, then the column number.)

$$\begin{bmatrix} 1 & -2 & 3 \\ 5 & -2 & -1 \end{bmatrix} \cdot \begin{bmatrix} 6 & -3 \\ 4 & 5 \\ 2 & 8 \end{bmatrix} = \begin{bmatrix} 1 \cdot 6 + (-2) \cdot 4 + 3 \cdot 2 & 1 \cdot (-3) + (-2) \cdot 5 + 3 \cdot 8 \\ 5 \cdot 6 + (-2) \cdot 4 + (-1) \cdot 2 & 5 \cdot (-3) + (-2) \cdot 5 + (-1) \cdot 8 \end{bmatrix}$$
$$= \begin{bmatrix} 4 & 11 \\ 20 & -33 \end{bmatrix}$$

Math.Content.N.VM.12

47. D: To transform a vector, $\vec{m}$, by matrix multiplication, start with the identity matrix of the appropriate dimension, in this case, $\boldsymbol{I} = \begin{bmatrix} 1 & 0 & 0 \\ 0 & 1 & 0 \\ 0 & 0 & 1 \end{bmatrix}$. To double the magnitude of $\overline{\mathrm{m}}$, multiply I by 2: $2 \cdot \begin{bmatrix} 1 & 0 & 0 \\ 0 & 1 & 0 \\ 0 & 0 & 1 \end{bmatrix} = \begin{bmatrix} 2 & 0 & 0 \\ 0 & 2 & 0 \\ 0 & 0 & 2 \end{bmatrix}$. To reflect across an axis, change the sign of the corresponding row. To reflect about the origin, change the sign in all rows: $\begin{bmatrix} -2 & 0 & 0 \\ 0 & -2 & 0 \\ 0 & 0 & -2 \end{bmatrix}$.

$$\begin{bmatrix} -2 & 0 & 0 \\ 0 & -2 & 0 \\ 0 & 0 & -2 \end{bmatrix} \cdot \begin{bmatrix} 3 \\ 2 \\ -1 \end{bmatrix} = \begin{bmatrix} -2 \cdot 3 + 0 + 0 \\ 0 + -2 \cdot 2 + 0 \\ 0 + 0 + -2 \cdot -1 \end{bmatrix} = \begin{bmatrix} -6 \\ -4 \\ 2 \end{bmatrix}$$

Math.Content.N.VM.10

48. B: The determinant of a 2x2 matrix is the product of the values on the main diagonal minus the product of the numbers on the other diagonal: if $\boldsymbol{M} = \begin{bmatrix} a & b \\ c & d \end{bmatrix}$, $det\boldsymbol{M} = ad - bc$. In this case: $det\boldsymbol{M} = \big(25 \cdot (-6)\big) - \big(15 \cdot (-10)\big) = -150 + 150 = 0$. Since the determinant equals 0, the inverse does not exist. (When the determinant is zero, the system of equations created to find the inverse evidence proportional coefficients and therefore have no solution.)

Math.Content.N.VM.12

49. C: The area of the parallelogram with legs $\vec{\mathrm{m}} = \begin{bmatrix} 5 \\ 7 \end{bmatrix}$ and $\vec{\mathrm{n}} = \begin{bmatrix} -3 \\ 10 \end{bmatrix}$ is equal to the determinant of the matrix formed by the two columns: $\begin{bmatrix} 5 & -3 \\ 7 & 10 \end{bmatrix}$.

$$det \begin{bmatrix} 5 & -3 \\ 7 & 10 \end{bmatrix} = 5 \cdot 10 - (-3) \cdot 7 = 50 + 21 = 71$$

Math.Content.N.VM.9

50. B: The associative property, $(A \cdot B) \cdot C = A \cdot (B \cdot C)$, is true for square matrices. The distributive property, $k(A \cdot B) = kA \cdot kB$, is also true for square matrices. The commutative property, $A \cdot B = B \cdot A$, is not true for matrix multiplication. The following is a counterexample:

$$\begin{bmatrix} 3 & 0 \\ 1 & 4 \end{bmatrix} \cdot \begin{bmatrix} 5 & 1 \\ 10 & 2 \end{bmatrix} = \begin{bmatrix} 15 & 3 \\ 45 & 9 \end{bmatrix} \neq \begin{bmatrix} 5 & 1 \\ 10 & 2 \end{bmatrix} \cdot \begin{bmatrix} 3 & 0 \\ 1 & 4 \end{bmatrix} = \begin{bmatrix} 16 & 4 \\ 32 & 8 \end{bmatrix}$$

How to Overcome Test Anxiety

Just the thought of taking a test is enough to make most people a little nervous. A test is an important event that can have a long-term impact on your future, so it's important to take it seriously and it's natural to feel anxious about performing well. But just because anxiety is normal, that doesn't mean that it's helpful in test taking, or that you should simply accept it as part of your life. Anxiety can have a variety of effects. These effects can be mild, like making you feel slightly nervous, or severe, like blocking your ability to focus or remember even a simple detail.

If you experience test anxiety—whether severe or mild—it's important to know how to beat it. To discover this, first you need to understand what causes test anxiety.

Causes of Test Anxiety

While we often think of anxiety as an uncontrollable emotional state, it can actually be caused by simple, practical things. One of the most common causes of test anxiety is that a person does not feel adequately prepared for their test. This feeling can be the result of many different issues such as poor study habits or lack of organization, but the most common culprit is time management. Starting to study too late, failing to organize your study time to cover all of the material, or being distracted while you study will mean that you're not well prepared for the test. This may lead to cramming the night before, which will cause you to be physically and mentally exhausted for the test. Poor time management also contributes to feelings of stress, fear, and hopelessness as you realize you are not well prepared but don't know what to do about it.

Other times, test anxiety is not related to your preparation for the test but comes from unresolved fear. This may be a past failure on a test, or poor performance on tests in general. It may come from comparing yourself to others who seem to be performing better or from the stress of living up to expectations. Anxiety may be driven by fears of the future—how failure on this test would affect your educational and career goals. These fears are often completely irrational, but they can still negatively impact your test performance.

Elements of Test Anxiety

As mentioned earlier, test anxiety is considered to be an emotional state, but it has physical and mental components as well. Sometimes you may not even realize that you are suffering from test anxiety until you notice the physical symptoms. These can include trembling hands, rapid heartbeat, sweating, nausea, and tense muscles. Extreme anxiety may lead to fainting or vomiting. Obviously, any of these symptoms can have a negative impact on testing. It is important to recognize them as soon as they begin to occur so that you can address the problem before it damages your performance.

The mental components of test anxiety include trouble focusing and inability to remember learned information. During a test, your mind is on high alert, which can help you recall information and stay focused for an extended period of time. However, anxiety interferes with your mind's natural processes, causing you to blank out, even on the questions you know well. The strain of testing during anxiety makes it difficult to stay focused, especially on a test that may take several hours. Extreme anxiety can take a huge mental toll, making it difficult not only to recall test information but even to understand the test questions or pull your thoughts together.

Effects of Test Anxiety

Test anxiety is like a disease—if left untreated, it will get progressively worse. Anxiety leads to poor performance, and this reinforces the feelings of fear and failure, which in turn lead to poor performances on subsequent tests. It can grow from a mild nervousness to a crippling condition. If allowed to progress, test anxiety can have a big impact on your schooling, and consequently on your future.

Test anxiety can spread to other parts of your life. Anxiety on tests can become anxiety in any stressful situation, and blanking on a test can turn into panicking in a job situation. But fortunately, you don't have to let anxiety rule your testing and determine your grades. There are a number of relatively simple steps you can take to move past anxiety and function normally on a test and in the rest of life.

Physical Steps for Beating Test Anxiety

While test anxiety is a serious problem, the good news is that it can be overcome. It doesn't have to control your ability to think and remember information. While it may take time, you can begin taking steps today to beat anxiety.

Just as your first hint that you may be struggling with anxiety comes from the physical symptoms, the first step to treating it is also physical. Rest is crucial for having a clear, strong mind. If you are tired, it is much easier to give in to anxiety. But if you establish good sleep habits, your body and mind will be ready to perform optimally, without the strain of exhaustion. Additionally, sleeping well helps you to retain information better, so you're more likely to recall the answers when you see the test questions.

Getting good sleep means more than going to bed on time. It's important to allow your brain time to relax. Take study breaks from time to time so it doesn't get overworked, and don't study right before bed. Take time to rest your mind before trying to rest your body, or you may find it difficult to fall asleep.

Along with sleep, other aspects of physical health are important in preparing for a test. Good nutrition is vital for good brain function. Sugary foods and drinks may give a burst of energy but this burst is followed by a crash, both physically and emotionally. Instead, fuel your body with protein and vitamin-rich foods.

Also, drink plenty of water. Dehydration can lead to headaches and exhaustion, especially if your brain is already under stress from the rigors of the test. Particularly if your test is a long one, drink water during the breaks. And if possible, take an energy-boosting snack to eat between sections.

Along with sleep and diet, a third important part of physical health is exercise. Maintaining a steady workout schedule is helpful, but even taking 5-minute study breaks to walk can help get your blood pumping faster and clear your head. Exercise also releases endorphins, which contribute to a positive feeling and can help combat test anxiety.

When you nurture your physical health, you are also contributing to your mental health. If your body is healthy, your mind is much more likely to be healthy as well. So take time to rest, nourish your body with healthy food and water, and get moving as much as possible. Taking these physical steps will make you stronger and more able to take the mental steps necessary to overcome test anxiety.

Mental Steps for Beating Test Anxiety

Working on the mental side of test anxiety can be more challenging, but as with the physical side, there are clear steps you can take to overcome it. As mentioned earlier, test anxiety often stems from lack of preparation, so the obvious solution is to prepare for the test. Effective studying may be the most important weapon you have for beating test anxiety, but you can and should employ several other mental tools to combat fear.

First, boost your confidence by reminding yourself of past success—tests or projects that you aced. If you're putting as much effort into preparing for this test as you did for those, there's no reason you should expect to fail here. Work hard to prepare; then trust your preparation.

Second, surround yourself with encouraging people. It can be helpful to find a study group, but be sure that the people you're around will encourage a positive attitude. If you spend time with others who are anxious or cynical, this will only contribute to your own anxiety. Look for others who are motivated to study hard from a desire to succeed, not from a fear of failure.

Third, reward yourself. A test is physically and mentally tiring, even without anxiety, and it can be helpful to have something to look forward to. Plan an activity following the test, regardless of the outcome, such as going to a movie or getting ice cream.

When you are taking the test, if you find yourself beginning to feel anxious, remind yourself that you know the material. Visualize successfully completing the test. Then take a few deep, relaxing breaths and return to it. Work through the questions carefully but with confidence, knowing that you are capable of succeeding.

Developing a healthy mental approach to test taking will also aid in other areas of life. Test anxiety affects more than just the actual test—it can be damaging to your mental health and even contribute to depression. It's important to beat test anxiety before it becomes a problem for more than testing.

Study Strategy

Being prepared for the test is necessary to combat anxiety, but what does being prepared look like? You may study for hours on end and still not feel prepared. What you need is a strategy for test prep. The next few pages outline our recommended steps to help you plan out and conquer the challenge of preparation.

Step 1: Scope Out the Test

Learn everything you can about the format (multiple choice, essay, etc.) and what will be on the test. Gather any study materials, course outlines, or sample exams that may be available. Not only will this help you to prepare, but knowing what to expect can help to alleviate test anxiety.

Step 2: Map Out the Material

Look through the textbook or study guide and make note of how many chapters or sections it has. Then divide these over the time you have. For example, if a book has 15 chapters and you have five days to study, you need to cover three chapters each day. Even better, if you have the time, leave an extra day at the end for overall review after you have gone through the material in depth.

If time is limited, you may need to prioritize the material. Look through it and make note of which sections you think you already have a good grasp on, and which need review. While you are studying, skim quickly through the familiar sections and take more time on the challenging parts.

Write out your plan so you don't get lost as you go. Having a written plan also helps you feel more in control of the study, so anxiety is less likely to arise from feeling overwhelmed at the amount to cover.

STEP 3: GATHER YOUR TOOLS

Decide what study method works best for you. Do you prefer to highlight in the book as you study and then go back over the highlighted portions? Or do you type out notes of the important information? Or is it helpful to make flashcards that you can carry with you? Assemble the pens, index cards, highlighters, post-it notes, and any other materials you may need so you won't be distracted by getting up to find things while you study.

If you're having a hard time retaining the information or organizing your notes, experiment with different methods. For example, try color-coding by subject with colored pens, highlighters, or post-it notes. If you learn better by hearing, try recording yourself reading your notes so you can listen while in the car, working out, or simply sitting at your desk. Ask a friend to quiz you from your flashcards, or try teaching someone the material to solidify it in your mind.

STEP 4: CREATE YOUR ENVIRONMENT

It's important to avoid distractions while you study. This includes both the obvious distractions like visitors and the subtle distractions like an uncomfortable chair (or a too-comfortable couch that makes you want to fall asleep). Set up the best study environment possible: good lighting and a comfortable work area. If background music helps you focus, you may want to turn it on, but otherwise keep the room quiet. If you are using a computer to take notes, be sure you don't have any other windows open, especially applications like social media, games, or anything else that could distract you. Silence your phone and turn off notifications. Be sure to keep water close by so you stay hydrated while you study (but avoid unhealthy drinks and snacks).

Also, take into account the best time of day to study. Are you freshest first thing in the morning? Try to set aside some time then to work through the material. Is your mind clearer in the afternoon or evening? Schedule your study session then. Another method is to study at the same time of day that you will take the test, so that your brain gets used to working on the material at that time and will be ready to focus at test time.

STEP 5: STUDY!

Once you have done all the study preparation, it's time to settle into the actual studying. Sit down, take a few moments to settle your mind so you can focus, and begin to follow your study plan. Don't give in to distractions or let yourself procrastinate. This is your time to prepare so you'll be ready to fearlessly approach the test. Make the most of the time and stay focused.

Of course, you don't want to burn out. If you study too long you may find that you're not retaining the information very well. Take regular study breaks. For example, taking five minutes out of every hour to walk briskly, breathing deeply and swinging your arms, can help your mind stay fresh.

As you get to the end of each chapter or section, it's a good idea to do a quick review. Remind yourself of what you learned and work on any difficult parts. When you feel that you've mastered the material, move on to the next part. At the end of your study session, briefly skim through your notes again.

But while review is helpful, cramming last minute is NOT. If at all possible, work ahead so that you won't need to fit all your study into the last day. Cramming overloads your brain with more information than it can process and retain, and your tired mind may struggle to recall even previously learned information when it is overwhelmed with last-minute study. Also, the urgent nature of cramming and the stress placed on your brain contribute to anxiety. You'll be more likely to go to the test feeling unprepared and having trouble thinking clearly.

So don't cram, and don't stay up late before the test, even just to review your notes at a leisurely pace. Your brain needs rest more than it needs to go over the information again. In fact, plan to finish your studies by noon or early afternoon the day before the test. Give your brain the rest of the day to relax or focus on other things, and get a good night's sleep. Then you will be fresh for the test and better able to recall what you've studied.

Step 6: Take a Practice Test

Many courses offer sample tests, either online or in the study materials. This is an excellent resource to check whether you have mastered the material, as well as to prepare for the test format and environment.

Check the test format ahead of time: the number of questions, the type (multiple choice, free response, etc.), and the time limit. Then create a plan for working through them. For example, if you have 30 minutes to take a 60-question test, your limit is 30 seconds per question. Spend less time on the questions you know well so that you can take more time on the difficult ones.

If you have time to take several practice tests, take the first one open book, with no time limit. Work through the questions at your own pace and make sure you fully understand them. Gradually work up to taking a test under test conditions: sit at a desk with all study materials put away and set a timer. Pace yourself to make sure you finish the test with time to spare and go back to check your answers if you have time.

After each test, check your answers. On the questions you missed, be sure you understand why you missed them. Did you misread the question (tests can use tricky wording)? Did you forget the information? Or was it something you hadn't learned? Go back and study any shaky areas that the practice tests reveal.

Taking these tests not only helps with your grade, but also aids in combating test anxiety. If you're already used to the test conditions, you're less likely to worry about it, and working through tests until you're scoring well gives you a confidence boost. Go through the practice tests until you feel comfortable, and then you can go into the test knowing that you're ready for it.

Test Tips

On test day, you should be confident, knowing that you've prepared well and are ready to answer the questions. But aside from preparation, there are several test day strategies you can employ to maximize your performance.

First, as stated before, get a good night's sleep the night before the test (and for several nights before that, if possible). Go into the test with a fresh, alert mind rather than staying up late to study.

Try not to change too much about your normal routine on the day of the test. It's important to eat a nutritious breakfast, but if you normally don't eat breakfast at all, consider eating just a protein bar. If you're a coffee drinker, go ahead and have your normal coffee. Just make sure you time it so that the caffeine doesn't wear off right in the middle of your test. Avoid sugary beverages, and drink enough water to stay hydrated but not so much that you need a restroom break 10 minutes into the test. If your test isn't first thing in the morning, consider going for a walk or doing a light workout before the test to get your blood flowing.

Allow yourself enough time to get ready, and leave for the test with plenty of time to spare so you won't have the anxiety of scrambling to arrive in time. Another reason to be early is to select a good seat. It's helpful to sit away from doors and windows, which can be distracting. Find a good seat, get out your supplies, and settle your mind before the test begins.

When the test begins, start by going over the instructions carefully, even if you already know what to expect. Make sure you avoid any careless mistakes by following the directions.

Then begin working through the questions, pacing yourself as you've practiced. If you're not sure on an answer, don't spend too much time on it, and don't let it shake your confidence. Either skip it and come back later, or eliminate as many wrong answers as possible and guess among the remaining ones. Don't dwell on these questions as you continue—put them out of your mind and focus on what lies ahead.

Be sure to read all of the answer choices, even if you're sure the first one is the right answer. Sometimes you'll find a better one if you keep reading. But don't second-guess yourself if you do immediately know the answer. Your gut instinct is usually right. Don't let test anxiety rob you of the information you know.

If you have time at the end of the test (and if the test format allows), go back and review your answers. Be cautious about changing any, since your first instinct tends to be correct, but make sure you didn't misread any of the questions or accidentally mark the wrong answer choice. Look over any you skipped and make an educated guess.

At the end, leave the test feeling confident. You've done your best, so don't waste time worrying about your performance or wishing you could change anything. Instead, celebrate the successful completion of this test. And finally, use this test to learn how to deal with anxiety even better next time.

Review Video: Test Anxiety
Visit mometrix.com/academy and enter code: 100340

Important Qualification

Not all anxiety is created equal. If your test anxiety is causing major issues in your life beyond the classroom or testing center, or if you are experiencing troubling physical symptoms related to your anxiety, it may be a sign of a serious physiological or psychological condition. If this sounds like your situation, we strongly encourage you to seek professional help.

How to Overcome Your Fear of Math

Not again. You're sitting in math class, look down at your test, and immediately start to panic. Your stomach is in knots, your heart is racing, and you break out in a cold sweat. You're staring at the paper, but everything looks like it's written in a foreign language. Even though you studied, you're blanking out on how to begin solving these problems.

Does this sound familiar? If so, then you're not alone! You may be like millions of other people who experience math anxiety. Anxiety about performing well in math is a common experience for students of all ages. In this article, we'll discuss what math anxiety is, common misconceptions about learning math, and tips and strategies for overcoming math anxiety.

What Is Math Anxiety?

Psychologist Mark H. Ashcraft explains math anxiety as a feeling of tension, apprehension, or fear that interferes with math performance. Having math anxiety negatively impacts people's beliefs about themselves and what they can achieve. It hinders achievement within the math classroom and affects the successful application of mathematics in the real world.

Symptoms and Signs of Math Anxiety

To overcome math anxiety, you must recognize its symptoms. Becoming aware of the signs of math anxiety is the first step in addressing and resolving these fears.

Negative Self-Talk

If you have math anxiety, you've most likely said at least one of these statements to yourself:

- "I hate math."
- "I'm not good at math."
- "I'm not a math person."

The way we speak to ourselves and think about ourselves matters. Our thoughts become our words, our words become our actions, and our actions become our habits. Thinking negatively about math creates a self-fulfilling prophecy. In other words, if you take an idea as a fact, then it will come true because your behaviors will align to match it.

Avoidance

Some people who are fearful or anxious about math will tend to avoid it altogether. Avoidance can manifest in the following ways:

- Lack of engagement with math content
- Not completing homework and other assignments
- Not asking for help when needed
- Skipping class
- Avoiding math-related courses and activities

Avoidance is one of the most harmful impacts of math anxiety. If you steer clear of math at all costs, then you can't set yourself up for the success you deserve.

Lack of Motivation

Students with math anxiety may experience a lack of motivation. They may struggle to find the incentive to get engaged with what they view as a frightening subject. These students are often overwhelmed, making it difficult for them to complete or even start math assignments.

Procrastination

Another symptom of math anxiety is procrastination. Students may voluntarily delay or postpone their classwork and assignments, even if they know there will be a negative consequence for doing so. Additionally, they may choose to wait until the last minute to start projects and homework, even when they know they need more time to put forth their best effort.

Physiological Reactions

Many people with a fear of math experience physiological side effects. These may include an increase in heart rate, sweatiness, shakiness, nausea, and irregular breathing. These symptoms make it difficult to focus on the math content, causing the student even more stress and fear.

Strong Emotional Responses

Math anxiety also affects people on an emotional level. Responding to math content with strong emotions such as panic, anger, or despair can be a sign of math anxiety.

Low Test Scores and Performance

Low achievement can be both a symptom and a cause of math anxiety. When someone does not take the steps needed to perform well on tests and assessments, they are less likely to pass. The more they perform poorly, the more they accept this poor performance as a fact that can't be changed.

Feeling Alone

People who experience math anxiety feel like they are the only ones struggling, even if the math they are working on is challenging to many people. Feeling isolated in what they perceive as failure can trigger tension or nervousness.

Feeling of Permanency

Math anxiety can feel very permanent. You may assume that you are naturally bad at math and always will be. Viewing math as a natural ability rather than a skill that can be learned causes people to believe that nothing will help them improve. They take their current math abilities as fact and assume that they can't be changed. As a result, they give up, stop trying to improve, and avoid engaging with math altogether.

Lack of Confidence

People with low self-confidence in math tend to feel awkward and incompetent when asked to solve a math problem. They don't feel comfortable taking chances or risks when problem-solving because they second-guess themselves and assume they are incorrect. They don't trust in their ability to learn the content and solve problems correctly.

Panic

A general sense of unexplained panic is also a sign of math anxiety. You may feel a sudden sense of fear that triggers physical reactions, even when there is no apparent reason for such a response.

Causes Of Math Anxiety

Math anxiety can start at a young age and may have one or more underlying causes. Common causes of math anxiety include the following:

The Attitude of Parents or Guardians

Parents often put pressure on their children to perform well in school. Although their intentions are usually good, this pressure can lead to anxiety, especially if the student is struggling with a subject or class.

Perhaps your parents or others in your life hold negative predispositions about math based on their own experiences. For instance, if your mother once claimed she was not good at math, then you might have incorrectly interpreted this as a predisposed trait that was passed down to you.

Teacher Influence

Students often pick up on their teachers' attitudes about the content being taught. If a teacher is happy and excited about math, students are more likely to mirror these emotions. However, if a teacher lacks enthusiasm or genuine interest, then students are more inclined to disengage.

Teachers have a responsibility to cultivate a welcoming classroom culture that is accepting of mistakes. When teachers blame students for not understanding a concept, they create a hostile classroom environment where mistakes are not tolerated. This tension increases student stress and anxiety, creating conditions that are not conducive to inquiry and learning. Instead, when teachers normalize mistakes as a natural part of the problem-solving process, they give their students the freedom to explore and grapple with the math content. In such an environment, students feel comfortable taking chances because they are not afraid of being wrong.

Students need teachers that can help when they're having problems understanding difficult concepts. In doing so, educators may need to change how they teach the content. Since different people have unique learning styles, it's the job of the teacher to adapt to the needs of each student. Additionally, teachers should encourage students to explore alternate problem-solving strategies, even if it's not the preferred method of the educator.

Fear of Being Wrong

Embarrassing situations can be traumatic, especially for young children and adolescents. These experiences can stay with people through their adult lives. Those with math anxiety may experience a fear of being wrong, especially in front of a group of peers. This fear can be paralyzing, interfering with the student's concentration and ability to focus on the problem at hand.

Timed Assessments

Timed assessments can help improve math fluency, but they often create unnecessary pressure for students to complete an unrealistic number of problems within a specified timeframe. Many studies have shown that timed assessments often result in increased levels of anxiety, reducing a student's overall competence and ability to problem-solve.

Debunking Math Myths

There are lots of myths about math that are related to the causes and development of math-related anxiety. Although these myths have been proven to be false, many people take them as fact. Let's go over a few of the most common myths about learning math.

Myth: Men Are Better at Math Than Women

Math has a reputation for being a male-dominant subject, but this doesn't mean that men are inherently better at math than women. Many famous mathematical discoveries have been made by women. Katherine Johnson, Dame Mary Lucy Cartwright, and Marjorie Lee Brown are just a few of the many famous women mathematicians. Expecting to be good or bad at math because of your gender sets you up for stress and confusion. Math is a skill that can be learned, just like cooking or riding a bike.

Myth: There Is Only One Good Way to Solve Math Problems

There are many ways to get the correct answer when it comes to math. No two people have the same brain, so everyone takes a slightly different approach to problem-solving. Moreover, there isn't one way of problem-solving that's superior to another. Your way of working through a problem might differ from someone else's, and that is okay. Math can be a highly individualized process, so the best method for you should be the one that makes you feel the most comfortable and makes the most sense to you.

Myth: Math Requires a Good Memory

For many years, mathematics was taught through memorization. However, learning in such a way hinders the development of critical thinking and conceptual understanding. These skill sets are much more valuable than basic memorization. For instance, you might be great at memorizing mathematical formulas, but if you don't understand what they mean, then you can't apply them to different scenarios in the real world. When a student is working from memory, they are limited in the strategies available to them to problem-solve. In other words, they assume there is only one correct way to do the math, which is the method they memorized. Having a variety of problem-solving options can help students figure out which method works best for them. Additionally, it provides students with a better understanding of how and why certain mathematical strategies work. While memorization can be helpful in some instances, it is not an absolute requirement for mathematicians.

Myth: Math Is Not Creative

Math requires imagination and intuition. Contrary to popular belief, it is a highly creative field. Mathematical creativity can help in developing new ways to think about and solve problems. Many people incorrectly assume that all things are either creative or analytical. However, this black-and-white view is limiting because the field of mathematics involves both creativity and logic.

Myth: Math Isn't Supposed To Be Fun

Whoever told you that math isn't supposed to be fun is a liar. There are tons of math-based activities and games that foster friendly competition and engagement. Math is often best learned through play, and lots of mobile apps and computer games exemplify this.

Additionally, math can be an exceptionally collaborative and social experience. Studying or working through problems with a friend often makes the process a lot more fun. The excitement and satisfaction of solving a difficult problem with others is quite rewarding. Math can be fun if you look for ways to make it more collaborative and enjoyable.

Myth: Not Everyone Is Capable of Learning Math

There's no such thing as a "math person." Although many people think that you're either good at math or you're not, this is simply not true. Everyone is capable of learning and applying mathematics. However, not everyone learns the same way. Since each person has a different learning style, the trick is to find the strategies and learning tools that work best for you. Some people learn best through hands-on experiences, and others find success through the use of visual aids. Others are auditory learners and learn best by hearing and listening. When people are overwhelmed or feel that math is too hard, it's often because they haven't found the learning strategy that works best for them.

Myth: Good Mathematicians Work Quickly and Never Make Mistakes

There is no prize for finishing first in math. It's not a race, and speed isn't a measure of your ability. Good mathematicians take their time to ensure their work is accurate. As you gain more experience and practice, you will naturally become faster and more confident.

Additionally, everyone makes mistakes, including good mathematicians. Mistakes are a normal part of the problem-solving process, and they're not a bad thing. The important thing is that we take the time to learn from our mistakes, understand where our misconceptions are, and move forward.

Myth: You Don't Need Math in the Real World

Our day-to-day lives are so infused with mathematical concepts that we often don't even realize when we're using math in the real world. In fact, most people tend to underestimate how much we do math in our everyday lives. It's involved in an enormous variety of daily activities such as shopping, baking, finances, and gardening, as well as in many careers, including architecture, nursing, design, and sales.

Tips and Strategies for Overcoming Math Anxiety

If your anxiety is getting in the way of your level of mathematical engagement, then there are lots of steps you can take. Check out the strategies below to start building confidence in math today.

Focus on Understanding, Not Memorization

Don't drive yourself crazy trying to memorize every single formula or mathematical process. Instead, shift your attention to understanding concepts. Those who prioritize memorization over conceptual understanding tend to have lower achievement levels in math. Students who memorize may be able to complete some math, but they don't understand the process well enough to apply it to different situations. Memorization comes with time and practice, but it won't help alleviate math anxiety. On the other hand, conceptual understanding will give you the building blocks of knowledge you need to build up your confidence.

Replace Negative Self-Talk with Positive Self-Talk

Start to notice how you think about yourself. Whenever you catch yourself thinking something negative, try replacing that thought with a positive affirmation. Instead of continuing the negative thought, pause to reframe the situation. For ideas on how to get started, take a look at the table below:

Instead of thinking...	Try thinking...
"I can't do this math." "I'm not a math person."	"I'm up for the challenge, and I'm training my brain in math."
"This problem is too hard."	"This problem is hard, so this might take some time and effort. I know I can do this."
"I give up."	"What strategies can help me solve this problem?"
"I made a mistake, so I'm not good at this."	"Everyone makes mistakes. Mistakes help me to grow and understand."
"I'll never be smart enough."	"I can figure this out, and I am smart enough."

Practice Mindfulness

Practicing mindfulness and focusing on your breathing can help alleviate some of the physical symptoms of math anxiety. By taking deep breaths, you can remind your nervous system that you are not in immediate danger. Doing so will reduce your heart rate and help with any irregular breathing or shakiness. Taking the edge off of the physiological effects of anxiety will clear your mind, allowing your brain to focus its energy on problem-solving.

Do Some Math Every Day

Think about learning math as if you were learning a foreign language. If you don't use it, you lose it. If you don't practice your math skills regularly, you'll have a harder time achieving comprehension and fluency. Set some amount of time aside each day, even if it's just for a few minutes, to practice. It might take some discipline to build a habit around this, but doing so will help increase your mathematical self-assurance.

Use All of Your Resources

Everyone has a different learning style, and there are plenty of resources out there to support all learners. When you get stuck on a math problem, think about the tools you have access to, and use them when applicable. Such resources may include flashcards, graphic organizers, study guides, interactive notebooks, and peer study groups. All of these are great tools to accommodate your individual learning style. Finding the tools and resources that work for your learning style will give you the confidence you need to succeed.

Realize That You Aren't Alone

Remind yourself that lots of other people struggle with math anxiety, including teachers, nurses, and even successful mathematicians. You aren't the only one who panics when faced with a new or challenging problem. It's probably much more common than you think. Realizing that you aren't alone in your experience can help put some distance between yourself and the emotions you feel about math. It also helps to normalize the anxiety and shift your perspective.

Ask Questions

If there's a concept you don't understand and you've tried everything you can, then it's okay to ask for help! You can always ask your teacher or professor for help. If you're not learning math in a traditional classroom, you may want to join a study group, work with a tutor, or talk to your friends. More often than not, you aren't the only one of your peers who needs clarity on a mathematical concept. Seeking understanding is a great way to increase self-confidence in math.

Remember That There's More Than One Way To Solve A Problem

Since everyone learns differently, it's best to focus on understanding a math problem with an approach that makes sense to you. If the way it's being taught is confusing to you, don't give up. Instead, work to understand the problem using a different technique. There's almost always more than one problem-solving method when it comes to math. Don't get stressed if one of them doesn't make sense to you. Instead, shift your focus to what does make sense. Chances are high that you know more than you think you do.

Visualization

Visualization is the process of creating images in your mind's eye. Picture yourself as a successful, confident mathematician. Think about how you would feel and how you would behave. What would your work area look like? How would you organize your belongings? The more you focus on something, the more likely you are to achieve it. Visualizing teaches your brain that you can achieve whatever it is that you want. Thinking about success in mathematics will lead to acting like a successful mathematician. This, in turn, leads to actual success.

Focus on the Easiest Problems First

To increase your confidence when working on a math test or assignment, try solving the easiest problems first. Doing so will remind you that you are successful in math and that you do have what it takes. This process will increase your belief in yourself, giving you the confidence you need to tackle more complex problems.

Find a Support Group

A study buddy, tutor, or peer group can go a long way in decreasing math-related anxiety. Such support systems offer lots of benefits, including a safe place to ask questions, additional practice with mathematical concepts, and an understanding of other problem-solving explanations that may work better for you. Equipping yourself with a support group is one of the fastest ways to eliminate math anxiety.

Reward Yourself for Working Hard

Recognize the amount of effort you're putting in to overcome your math anxiety. It's not an easy task, so you deserve acknowledgement. Surround yourself with people who will provide you with the positive reinforcement you deserve.

Remember, You Can Do This!

Conquering a fear of math can be challenging, but there are lots of strategies that can help you out. Your own beliefs about your mathematical capabilities can limit your potential. Working toward a growth mindset can have a tremendous impact on decreasing math-related anxiety and building confidence. By knowing the symptoms of math anxiety and recognizing common misconceptions about learning math, you can develop a plan to address your fear of math. Utilizing the strategies discussed can help you overcome this anxiety and build the confidence you need to succeed.

Thank You

We at Mometrix would like to extend our heartfelt thanks to you, our friend and patron, for allowing us to play a part in your journey. It is a privilege to serve people from all walks of life who are unified in their commitment to building the best future they can for themselves.

The preparation you devote to these important testing milestones may be the most valuable educational opportunity you have for making a real difference in your life. We encourage you to put your heart into it—that feeling of succeeding, overcoming, and yes, conquering will be well worth the hours you've invested.

We want to hear your story, your struggles and your successes, and if you see any opportunities for us to improve our materials so we can help others even more effectively in the future, please share that with us as well. **The team at Mometrix would be absolutely thrilled to hear from you!** So please, send us an email (support@mometrix.com) and let's stay in touch.

If you'd like some additional help, check out these other resources we offer for your exam: http://MometrixFlashcards.com/CCSS

Online Resources

Due to our efforts to try to keep this book to a manageable length, we've created a link that will give you access to all of your online resources:

mometrix.com/resources719/cchsnumquant

It's Your Moment, Let's Celebrate It!

Share your story @mometrixtestpreparation